CODE OPTIMIZATION TECHNIQUES FOR EMBEDDED PROCESSORS

Code Optimization Techniques for Embedded Processors

Methods, Algorithms, and Tools

by

Rainer Leupers

University of Dortmund

KLUWER ACADEMIC PUBLISHERS
BOSTON / DORDRECHT / LONDON

A C.I.P. Catalogue record for this book is available from the Library of Congress.

ISBN 978-1-4419-5010-9

Published by Kluwer Academic Publishers,
P.O. Box 17, 3300 AA Dordrecht, The Netherlands.

Sold and distributed in North, Central and South America
by Kluwer Academic Publishers,
101 Philip Drive, Norwell, MA 02061, U.S.A.

In all other countries, sold and distributed
by Kluwer Academic Publishers,
P.O. Box 322, 3300 AH Dordrecht, The Netherlands.

Printed on acid-free paper

This printing is a digital duplication of the original edition.

Printed in the United States of America

Contents

Foreword

The building blocks of today's and future embedded systems are complex intellectual property components, or cores, many of which are programmable processors. Traditionally, these embedded processors mostly have been programmed in assembly languages due to efficiency reasons. This implies time-consuming programming, extensive debugging, and low code portability. The requirements of short time-to-market and dependability of embedded systems are obviously much better met by using high-level language (e.g. C) compilers instead of assembly. However, the use of C compilers frequently incurs a code quality overhead as compared to manually written assembly programs. Due to the need for efficient embedded systems, this overhead must be very low in order to make compilers useful in practice. In turn, this requires new compiler techniques that take the specific constraints in embedded system design into account. An example are the specialized architectures of recent DSP and multimedia processors, which are not yet sufficiently exploited by existing compilers.

This book concentrates on new code optimization techniques dedicated to several classes of embedded processors. It summarizes the state-of-the-art in code generation for embedded processors and describes results of my research work at the Embedded Systems Group of the University of Dortmund between 1997 and 2000. In an earlier book on code generation for embedded processors [Leup97], emphasis has been on retargetability issues, including methods for automatically generating compilers from processor descriptions. In contrast, this book is focused on code optimization, where specific processor architectural features are exploited in order to generate efficient code. Nevertheless, the presented techniques are still machine-independent, and thus retargetable, to a certain extent.

Contributions to the results described in this book have been made by several people, whose help is gratefully acknowledged. The offset assignment technique from chapter 2 has been implemented by Fabian David as a part of his Diploma thesis, while the array optimization described in the same chapter has been developed in cooperation with Prof. Anupam Basu (IIT Kharagpur, India) during his visit at Dortmund University. The students Steffen Dingel and Daniel Kotte (Dortmund) contributed significantly to the original implementation of the LANCE compiler system described in chapter 8. Further software modules have been developed by the Indian guest students Vivek Haldar, Rahul Gupta, and Amit Manjhi. The research described in this book has been financially supported by Dr. Rajiv Jain's department at Agilent Technologies (Westlake Village, USA). Finally, I would like to thank the head of our Embedded Systems Group, Prof. Peter Marwedel, for his support, my colleague Jens Wagner for comments on the manuscript, Dr. Ashok Sudarsanam (Princeton University) for making the Olive code generation tool available, and all reviewers of this book for their efforts.

I dedicate this book to my father Hans Joachim Leupers: Thank you for sharing your knowledge with me and for many early morning trips to the airport !

Dortmund, July 2000   Rainer Leupers

Chapter 1

INTRODUCTION

According to the 1998 SIA roadmap [SIA98], integrated circuits (ICs) at the end of this decade will typically have more than 500 million transistors, an integration scale more than one order of magnitude higher than in today's microelectronics. For the high performance processor segment, even larger integration scales have been predicted, with processor performance reaching 3 million MIPS by the year 2010 [Tech00].

This rapid progress in semiconductor technology together with the corresponding increase in IC complexity create new demands on design automation tools. One key aspect is the abstraction level, at which electronic systems are specified. Today, design engineers mostly use hardware description languages such as VHDL [LSU89] and Verilog [ThMo91] to specify hardware structures at the register-transfer level, while using libraries of predefined macrocells like memories or multipliers. By means of available CAD tools, such specifications can be automatically mapped to IC mask layouts.

In certain application domains a more advanced behavioral specification style has become common, where systems are described by (and are synthesized from) their functionality rather than their structure. This approach, called *behavioral* or *high-level synthesis* [GDWL92], is commonly used in the area of digital signal processing, where synthesis starts from data flow specifications [BML96], and control-dominated applications, which are synthesized from extended finite automata models [DrHa89].

Still, this design methodology is not sufficient to ensure high productivity for future complex systems. Correct designs with a short time-to-market can only be achieved if designers no longer need to deal with "low-level" components like registers, adders, multiplexers, or even logic gates. In other words, specifications need to be raised to a higher abstraction level which permits a higher degree of component reuse. Design systems focusing on reuse of com-

1

plex components have become available only recently (e.g. the tool suite from Y Explorations [YXI00]).

High-level specifications are supported by the availability of *cores*, sometimes more generally denoted as *intellectual property*. Cores can be regarded as complex macrocells, that implement small and frequently required application subroutines (e.g. audio data compression) or even complete programmable processors. Cores are generally available in the form of synthesizable VHDL or Verilog models, or as predefined IC mask layout cells. In either case, designers may reuse cores like library components in order to achieve short design times. An additional advantage of using programmable processor cores as building blocks is their high flexibility. This allows for easier upgrading and debugging as compared to non-programmable hardware.

Since programmable processor cores are increasingly being used as components of systems-on-a-chip [PCL+96], design automation environments for electronic systems are no longer restricted to hardware synthesis and validation, but they also need to incorporate software development tools for processor cores. In this book we focus on *compilers*, and we take into account the special demands on compilers in the design of *embedded systems*. Embedded systems [GVNG94] are a special class of electronic systems that can be briefly characterized by the following terms:

Application-specific: The application of an embedded system is fixed in advance. Therefore, the design can be tuned for the given application, and there is no need to implement standard interfaces to peripheral devices like keyboards and hard disk drives. Unlike a general-purpose computer, for instance a PC, an embedded system is usually not reprogrammed or upgraded during its lifetime. Typical application domains of embedded systems include telecommunication (e.g. mobile phones), automotive electronics (e.g. engine control), and consumer electronics (e.g. digital cameras).

Reactive: An embedded system typically reacts on events coming from its environment, in which case it performs some kind of processing and produces new events. Very often, this processing has to be performed under real-time constraints, i.e., the system has to react at a speed determined by its environment. An airbag controller, for instance, has to trigger very quickly after a crash has been sensed, and a digital answering machine has to sample and encode speech signals at a predefined rate. This is also in contrast to general-purpose systems, where a general goal is to *maximize* performance. Instead, for a reactive system, the performance normally is a *constraint*.

Efficient: For a *fixed* application, such as text processing, a PC is definitely not an efficient solution. But, since PCs are intended to run numerous different applications, vendors and customers do not need to care about this.

The situation is very different for embedded systems. For instance, smaller silicon area directly translates to lower production costs of a system-on-a-chip, and lower power consumption of a mobile device ensures a longer battery lifetime. Therefore, non-specialized components such as a Pentium CPU, are generally not a good choice in the design of embedded systems. Instead, in very cost-sensitive embedded system domains (such as consumer electronics) and in domains with special environmental constraints (such as battery power supply) the efficiency of a design is extremely important for market success.

application-specific:	reactive:	efficient:
programmability restricted to number directory, ring melody, ...	idle most of the time, real-time processing when active	small size, low power consumption, low fabrication costs

Figure 1.1.   Example of an embedded system

Fig. 1.1 illustrates these characteristics for a widespread type of embedded systems. Sometimes, embedded systems have to meet further requirements, such as high dependability (for safety-critical applications), user-friendliness (for customers not used to operating computers), and low physical volume (for portable applications). However, as these requirements are not directly related to compiler techniques, we do not treat them in further detail here.

The design of embedded systems is typically considered as a *hardware-software codesign* problem [KAJW96]. Starting from a formal system speci-fication, the system to be designed is first partitioned into hardware and soft-ware components with respect to given constraints and objectives, dependent on the application domain [Niem98]. The hardware components can then be synthesized, while the software components are supposed to be executed on a programmable processor. In today's embedded systems, the amount of soft-ware mostly exceeds the amount of hardware [Arno00], which stresses the importance of software compilers in a hardware/software codesign process.

Processors used in embedded systems are commonly called *embedded processors*. In the following, we will give an overview of different families of embedded processors, and we will show to which extent the special characteristics of embedded systems determine the demands on compilers.

1.   EMBEDDED PROCESSOR CLASSES

The choice of a suitable processor to be integrated into an embedded system mainly depends on the application domain of that system. Some applications are arithmetic-intensive, while others are control-intensive or very sensitive to fabrication costs or power consumption. Consequently, in order to allow for efficient designs, there exist different classes of embedded processors, whose architectures are briefly characterized in this section.

1.1   MICROCONTROLLERS

Microcontrollers are relatively slow, but very area-efficient processors for control-intensive applications. They usually show a microprogrammed CISC (*complex instruction set computer*) architecture, which means that the number of clock cycles needed to execute different instructions show a large variance. Microcontrollers come with very limited computational and storage resources, and the data path word length is relatively small (typically 8 or 16 bits). On the other hand, their complex instruction set provides a convenient programming interface, with instructions for multiplication and division and a variety of memory addressing modes. This results in very dense code. Due to their control-oriented application domain, microcontrollers typically provide a rich set of instructions for bit-level data manipulation, as well as peripheral components like timers or serial I/O ports. Frequently, simple processors that earlier served as CPUs in computer systems (such as the 6502 [Nabe00], which has been a popular CPU for home computers[1] in the eighties) are nowadays being reused in a customized form as microcontrollers for embedded systems.

1.2   RISC PROCESSORS

RISC (*reduced instruction set computer*) processors evolved from CISC architectures due to the observation, that many complex instructions are executed only rarely in practical applications [HePa90]. Therefore, as a counterpart to CISCs, RISC processors offer only a very basic set of instructions, which can be executed at very high speed, though. Since all instructions of a RISC typically have the same size and require the same number of clock cycles, the execution

---

[1] Comparing computer games that ran on a 1 MHz 6502 in a Commodore C64 to contemporary ones running on a 500 Mhz PC today indicates that the 6502 is among the most efficiently programmed processors that have ever been used.

of instructions can be pipelined in order to achieve a higher throughput than in a purely sequential execution mechanism.

Another characteristic of RISCs is a load-store architecture with a large number of general-purpose registers in order to reduce the number of memory accesses in a machine program. In fact, a careful assignment of program variables to registers is among the most important optimizations a compiler for RISCs should perform. Still, for a fixed application, the code size for a RISC generally exceeds the code size for a CISC processor. A popular member of the RISC processor class in the context of embedded systems is the ARM RISC core family [ARM00]. The ARM7 core is a 32-bit RISC processor, whose instruction format can be switched between a length of 16 or 32. Its architecture has been specifically designed for low power consumption, typically in the order of only 100 mW, which makes the ARM suitable for portable systems with battery supply.

1.3   DIGITAL SIGNAL PROCESSORS

Digital signal processors (DSPs) have been designed for arithmetic-intensive signal processing applications [Lee88]. Consequently, their instruction sets are tuned for fast execution of algorithms like digital filtering and Fast Fourier Transform. This is supported by special hardware components, e.g., hardware multipliers and dedicated address generation units. In order to allow for efficient signal processing, DSPs usually show a certain degree of *instruction-level parallelism*, i.e., certain combinations of instructions can be executed in parallel in each instruction cycle.

Another typical feature of DSPs are *special-purpose registers*, e.g. a dedicated accumulator register. Unlike registers in RISCs, the use of special-purpose registers is dependent on the instruction context: Instructions must read their arguments from certain registers and also must write the result to a certain register. This leads to shorter combinational delays in the data paths and requires less bits for instruction encoding.

Additionally, DSPs may be operated in special arithmetic modes. In saturation mode, for instance, positive computation results exceeding the register word length do not overflow into the negative number domain like in usual twos-complement arithmetic, but all overflown values are converted into the largest representable number. Due to the irregularities in the processor architecture, as compared to other processor classes, compiler construction for DSPs is more difficult, at least when efficient code has to be generated. The market leader in DSPs is Texas Instruments, which offer a large variety of "off-the-shelf" DSPs for different application domains, such as audio or video signal processing. Other important semiconductor vendors offering DSPs include Motorola, Analog Devices, and NEC.

1.4   MULTIMEDIA PROCESSORS

Multimedia processors are relatively new on the semiconductor market, and they are architecturally related to RISCs and DSPs. However, they are intended to serve as uniform platforms for multimedia applications, which typically include audio, image, or video signal processing, but also for general-purpose "glue code". The architecture of multimedia processors follows the VLIW (*very long instruction word*) paradigm: Different functional units can operate in parallel and are controlled by separate fields in the instruction word. In contrast to *superscalar* processors, only the compiler is responsible for exploiting potential parallelism in machine programs.

In a detailed study, Fisher et al. [FDF98a] point out why VLIW might be *the* architecture of choice for digital and multimedia signal processing in the future. Such processors show a high degree of instruction-level parallelism and thereby allow for very high performance. In total, the architecture is more regular than in DSPs. For instance, there are general-purpose registers like in a RISC. Since VLIW and general-purpose registers are well-known concepts in compiler construction, many existing compiler techniques can be reused for multimedia processors. On the other hand, there are features not present in other processor classes. These include *conditional instructions* for fast execution of if-then-else statements, as well as SIMD (*single instruction multiple data*) instructions for higher resource utilization in computations on data with short word lengths.

Examples for multimedia processors are the Texas Instruments C6201 [TI00] (up to 8 parallel instructions per cycle) and the Philips Trimedia TM1000 [Phil00] (up to 5 parallel instructions). Especially for this class of embedded processors compiler support is very urgent, because the solution space for mapping a source program to an assembly program is so large that a human assembly programmer, usually not used to thinking in terms of parallel (and possibly also conditional or SIMD) instructions, will most probably not be able to produce code of high quality within a reasonable amount of time.

1.5   APPLICATION-SPECIFIC PROCESSORS

The processor classes mentioned so far are domain-specific. They are tuned for a certain application domain, but not for for the given application itself. In contrast, *application-specific instruction set processors* (ASIPs) are a compromise between domain-specific processors and non-programmable ASICs. ASIPs are still programmable, but they serve only a very narrow range of applications. Sometimes, ASIPs can be parameterized: In this case, the basic architecture of an ASIP is fixed, but it can be customized for a given application by setting a number of different parameters. In this way, word lengths may be adjusted to the required precision, register files may be sized, and available special hardware components, e.g. hardware loop support, may or may not be included. Since these parameters are mostly orthogonal to each other, a large number of different configurations of a single ASIP may be available. Consequently, ASIPs are very efficient, but a large number of different compilers would normally be required.

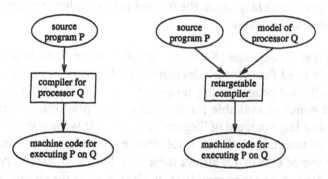

Figure 1.2.   Regular versus retargetable compilation

In order to avoid this effort, *retargetable* compilers can be used, capable of generating code for any particular ASIP configuration. In addition to the source code to be compiled, retargetable compilers read a model of the target processor as an input (fig. 1.2). However, retargetability is not easy to implement, and usually affects code quality. Therefore, such compilers are still rarely used in practice. An example for a parameterizable ASIP is the AMS Gepard DSP core [AMS00], whose parameters include the following: data word length (8 to 64 bits), program address width (8 to 19 bits), number of accumulators (2 to 4), and number of index registers (8 or 16). A very different ASIP architecture is the M3, a high-performance DSP for mobile applications [FWD+98]. The M3 shows a scalable SIMD architecture with up to 16 parallel arithmetic units with identical local instruction sets.

Note that processors frequently do not strictly fall into only one class. There are, for instance, new RISC-like microcontrollers, RISC processors may provide some DSP instructions, and some off-the-shelf DSPs are also available in a customizable form, so as to serve as ASIPs.

Different processor classes demand for different compiler techniques. In this book, we focus on two processor classes which are particularly challenging for compiler technology: DSPs and multimedia processors.

2.   DEMANDS ON COMPILERS
2.1   PROGRAMMING LANGUAGE

Due to the need for very efficient machine code and the lack of compilers generating efficient code, assembly-level software development for embedded processors is still very common, in particular in the DSP area. However, as both embedded applications and embedded processors are getting more and more complex, there is a need to replace assembly programming by the use of high-level language compilers. So the question arises, which language is suitable for programming embedded processors. Among the large number of available programming languages, the following have gained importance in the design of embedded systems:

C: The original C language [KeRi88], although more than 20 years old, is still widely used for software development. One main reason is that C is a very well-tried programming language, which is comparatively easy to learn and which is available for most computer platforms. Additionally, there exists a large amount of "legacy" source code written in C, which still needs to be maintained. From a software engineering viewpoint, a major disadvantage of C is that it allows for a very low-level, assembly-like, and machine-dependent programming style. However, in the context of embedded systems, it is exactly this feature which makes C attractive. Features like direct access to physical memory via pointers, post-increment of variables, and "dirty" type casts are frequently needed for developing system software in a high-level language. In addition, the concept of separately compilable modules in C allows for an easy inclusion of machine-specific functions written in assembly.

C++: The object-oriented variant of the C language [Stro87] has been developed in an effort to combine the advantages of object-oriented software development and the flexibility offered by C. Enabled by the decision to include C as a subset of C++, this language is step-by-step replacing the original C language. C++ strongly encourages a high-level programming style, based on class libraries for primitive data structures, but on the other hand enables the reuse of existing code. In the context of embedded systems, C++ also bears some disadvantages. Language features like the throw-catch mechanism for exception handling and templates for typeless function parameters are comfortable from a programmer's viewpoint, but they cause an overhead in size and performance of the compiled code. Therefore, the "Embedded C++ Technical Committee" has agreed upon a simplified C++ standard, called EC++ [Plau00], suitable for embedded software development. Development tools supporting EC++ are already available, but EC++ is not yet a widespread language.

Java: Java is another object-oriented language, primarily developed for platform-independent applications than can be run via the internet. Consequently, it is completely machine-independent and it does not support hardware oriented language elements like pointers. Java allows for very compact code, since it is usually compiled into a *byte code representation*, an intermediate format supposed to be interpreted by a stack-based Java Virtual Machine [LiYe97]. In the context of embedded system design, Java appears to be attractive due to advanced language features like automatic garbage collection and multithreading. Approaches to system design based on Java are described, e.g. in [RoWe97, PSLM+98, KuRo00]. On the other hand, the Java language features require a special runtime system and (unless Java is directly compiled into native machine code) also the byte code interpreter itself must be present "on-chip" in order to execute the byte code. In order to meet the efficiency requirements of embedded systems, restricted Java dialects, such as Embedded Java [Sun00] are being specified.

DFL: The data flow language DFL [Hilf85, Ment93] has been developed for the specification of DSP systems. The syntax resembles C, but there are special language constructs useful for describing DSP algorithms. These include the specification of saturation and rounding behavior for arithmetic operations, the concept of *delayed signals* (needed for digital filters), and the capability of textually capturing *signal flow graphs*, a common representation of DSP algorithms. DFL has been supported by commercial design automation tools for some time. However, even though it allows for a rather comfortable programming style in the DSP area, DFL has not really been a success on the tool market. Obviously, the main reason was that it is rather difficult to establish new language standards for narrow application domains.

In summary, there is still no ideal language for programming embedded processors, and there will probably never be such a language. Many of the code optimization techniques presented in the following chapters are actually independent of a concrete source language, but from time to time we need to refer to concrete source language constructs.

In this book, we will use C as a basis for compiler techniques. The reason is that C still is among the most important programming languages, especially in the embedded market segment, and that the development effort for C compilers is lower than for other languages mentioned above. C reference code exists for many standards in the DSP area (e.g. for voice compression [Lee00]).

In addition, C is an important *exchange format* in system-level design tools. For instance, C code can be automatically generated by tools from more abstract specification formalisms, such as Statecharts [Hare87]. Also in many Hardware/Software Codesign tools, e.g. Cosyma [EHB93], C programs are generated as an intermediate format for software components in embedded systems. Therefore, even if systems are specified at a higher level of abstraction than in programming languages, good C compilers are in practice still required to efficiently map software into machine code for embedded processors. Finally, recent efforts in the context of the SystemC initiative [SC00, Arno00] and the design of the SpecC language [GZD+00] aim at introducing C/C++ also as a platform for *hardware* modeling. This means that the C language (with the C++ extensions), although originally developed for a very different purpose, might end up as a unified hardware/software specification language in the future.

2.2   PERFORMANCE

As already mentioned, an important characteristic of embedded systems is reactivity, frequently combined with real-time constraints. That is, events coming in from the system environment have to be processed within a fixed amount of time. Meeting such constraints would frequently not be a problem when using high-performance general-purpose processors. However, due to the need for efficiency, designers can usually not afford to implement very specific applications for instance on a 800 MHz Pentium processor. Instead, typically domain-specific processors are used whose performance, if programmed efficiently, is merely sufficient to meet the constraints.

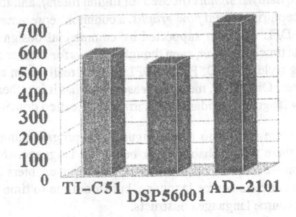

Figure 1.3. DSPStone benchmarking: Execution cycle overhead (percent, relative to assembly code) of compiled code for a DSP application (ADPCM transcoder) and three different processors

No compiler can generate provably optimal code for arbitrary source programs[2]. Therefore, a common reference for "efficient" machine code is assembly code, manually written by expert programmers. Such code can be regarded as close-to-optimal. On the other hand it is well-known that compiler-generated code usually shows an overhead, both in code size and performance, as compared to equivalent assembly code. In order to allow for the use in embedded software development, this overhead needs to be very small. For a given cost-sensitive application, which could be mapped to a 10 MHz processor when

---

[2]Otherwise, the *halting problem* would be decidable [Appe98].

programming in assembly, nobody would be willing to switch to a more expensive 20 MHz processor just because the compiler generates code with only half the performance of the assembly code. The compiler would simply not be used in this case.

This problem is observed especially in the DSP area. The DSPStone benchmarking project [ZVSM94] showed that for a number of widespread DSPs the performance of the corresponding C compilers, when comparing compiled code to reference assembly code, is so poor that compilers can hardly be used. At least this holds for time-critical software. For DSPs, the performance overhead of compiled code versus assembly reference may amount to as much as several hundred % (fig. 1.3). Recent updates of DSPStone for new DSPs [RLW99] as well as further experimental studies [Levy97, CWKL+99] indicate that, as far as commercial compilers are concerned, this situation has not changed significantly since the middle of the 90s. There is still a lack of compilers that efficiently make use of the special hardware features of DSPs. Even though a number of DSP-specific code optimization techniques have been published, more such techniques are definitely required in order to enable the step from assembly to C programming of DSPs. In addition, new processors with new hardware features are continuously being developed (e.g. multimedia processors), for which satisfactory compiler technology is not yet available.

A related area is the estimation of the *worst-case execution time* of embedded software, which is important for real-time software. Such estimations are required in early phases of a system design process, as well as for validation purposes after a design is finished. Estimation techniques that also take caching effects into account are described in [FMW97, LMW99].

2.3 CODE SIZE

Even though code size is not directly related to performance, the DSPStone project [ZVSM94] showed similar figures for the code size overhead of compiled code as for the performance overhead. In embedded systems-on-a-chip, where program memories are integrated with cores, a larger code size immediately implies a lower chip yield and thus higher costs. Thus, also the code size overhead of compiled code needs to be very low as compared to assembly code. This is in sharp contrast to general-purpose systems, where application programmers today usually do not need to care much about code size. During the discussion of concrete code optimization techniques in the following chapters we will see, that some optimizations have a positive effect both on code size and performance, while in other cases there is a trade-off between the two optimization goals. Eventually, it depends on the concrete application and technology at hand, which optimization goal should have priority.

A different approach to reduce program memory size for embedded systems is *code compression*. The main idea is to compute a compact encoding for a *fixed* binary program and to replace that program on-chip by the encoded program and a corresponding runtime decoding hardware. Recent techniques are described in [LeWo98, LDK99, CAP99], while [Wiel00] provides a comprehensive survey. Since code compression is only weakly related to code generation, this technique can be considered complementary to code size minimization in compilers.

2.4   POWER CONSUMPTION

One optimization goal that has received significant attention in embedded system design recently is low power consumption [RaPe96, RJD98, BeDe98, RRD99]. In some cases, for instance for high-performance processors, minimizing the *peak* power consumption is of main interest in order to avoid chip damages due to too high currents. In embedded systems, many of which are mobile and thus come with a battery supply, minimizing the *energy* consumption is frequently of great importance. A number of hardware design techniques for low power are already available, and there are embedded processors with special instructions for switching into a low power mode until an interrupt is received.

However, also the software of embedded systems has a large impact on power consumption. For example, Siemens recently announced that the standby power consumption of their C25 mobile phone has been reduced by 60 % through a pure software modification. Very few research works have focused on the actual role of compilers for low power. On one hand it is obvious that performance optimization of compiled code may be in itself an important contribution to low power, since under a given real-time constraint a more efficiently programmed processor can be operated at a lower clock frequency. However, from a general viewpoint, a compiler can do more for low power. For instance, it has been shown that special code optimizations that are not related to or even contrary to the usual objectives, performance and code size, may reduce the power consumption of a fixed application on a fixed processor [LTMF97, Gebo97b]. This opens new opportunities for reducing the power consumption of embedded systems. Both existing code optimization techniques and novel techniques (e.g. based on alternative data and instruction encodings) should be investigated. This is the subject of ongoing research [MLF98].

2.5   CORRECTNESS

Like any other software tool, compilers for embedded processors should be correct. This means, a compiled machine program should be functionally equivalent to its source program. Since compilers are basic tools for developing other software, compilers are expected to be even "more correct" (i.e., they should have less bugs) than other tools.

There are different ways of maximizing compiler correctness. One way is verification, i.e, the correctness is formally proven. There has been some success in verifying parts of a compiler, e.g. [GHZG99]. However, as a compiler is a quite complex software whose size may well exceed 100,000 lines of source code, complete verification of a C compiler currently seems out of scope due to too high runtime requirements.

A more practical, yet less reliable, way of compiler validation is simulation. In almost all cases, compilers for embedded processors are *cross-compilers*, i.e., they generate code for a target machine different from the compiler host machine. Consequently, simulators are needed that emulate the target machine by executing host instructions. Using a set of test programs and test data, one can identify bugs in the compiler by compiling and simulating the target machine code and comparing the program output to the expected output. A number of C validation suites for this purpose are commercially available (e.g. the Plum Hall Validation Suite [Plum00].

A difficult problem in the context of processor simulation is the *simulation speed*. Many available processor simulators based on the classical *interpretive* approach, where target machine instructions are interpreted by host instructions step-by-step, are extremely slow. In experiments for the Texas Instruments C6201 processor, we found that the corresponding commercial instruction set simulator has a speed in the order of 3750 simulated instruction cycles per host CPU second. The usual clock frequency of the C6201 is 200 MHz. This means, that the simulation of a program on the host takes more than 50,000 times longer than the execution of the program on the real machine. To exemplify this, the simulation of a single second of real-time processing takes more than 14 hours of simulation time. In order to eliminate this bottleneck, new instruction set simulation techniques, based on the *compiled simulation* approach [ZTM95, HRR+97, ZhGa99], have been introduced recently. Their key idea is to move a large part of the simulation effort from simulator runtime to simulator generation time.

The construction of correct compilers is partially supported by *generator technology*. The main idea is to automatically generate compiler source code from more abstract specifications, which in turn greatly reduces the number of bugs in the compiler. Generators for source language frontends include the well-known UNIX tools lex and yacc, which emit source code for parsing context-free languages specified by a grammar. Using generators is also common for implementing *program analyzers* [AlMa95], *optimizing transformations* [Assm98], and *code selectors* [AGT89], which play a key role for efficient code generation. Unfortunately, only a small fraction of the tasks to be performed by a compiler are covered by generators, especially in the case of embedded processors with unusual hardware features. However, in this book we show, that we can sometimes take advantage of standard generators even for very special code optimizations, by appropriately embedding the generated compiler parts into custom techniques.

In spite of the availability of generators, the validation of compilers remains a difficult problem, especially in the case of cross-compilers. In chapter 8, where we discuss some implementation issues, we will show a simple yet effective approach, that supports the validation of at least the machine-independent parts of a compiler.

2.6   RETARGETABILITY

As already mentioned, retargetable compilers are very useful in code generation for ASIPs, since they avoid the need to write different compilers for all different configurations of an ASIP. In addition, retargetable compilers can help to determine the best configuration of an ASIP for a given application, since they allow to study the effect of changes in the processor architecture on the performance or size of the generated code. Essentially, this permits to make hardware-software trade-offs during system design. This concept is already finding its way into industrial practice. For instance, Tensilica [Tens00] offers a WWW-based configuration mechanism for customizing a proprietary RISC-based ASIP architecture by numerous parameters. For each configuration, hardware description language (HDL) synthesis and simulation models as well as a generated GNU-based C++ compiler can be downloaded and locally evaluated. In this way, the adequate processor configuration for the intended application can be semi-automatically determined within several iterations.

Alternatively, also retargetable estimation techniques can be used to determine a good application-specific architecture [GNR00]. This avoids the need of constructing complete compilers, however at the expense of lower accuracy.

In order to be retargetable, a compiler – to a certain extent – has to be machine-independent. The compiler can be adapted to a certain target machine by writing custom machine-specific compiler components, or by providing the compiler with a model of the target machine, for which it has to generate code. Unfortunately, retargetability inherently tends to compromise code efficiency. This is due to the fact that the fewer assumptions the compiler can make about the target machine, the less machine-specific hardware features can be exploited to generate efficient code.

Thus, for embedded system designs not based on ASIPs but on "off-the-shelf" processors, code optimizations should have priority over retargetability in order to meet the goal of efficient implementations. This situation could change in the future, if the use of ASIPs becomes more common. In fact, retargetable compilers have been built that generate reasonably good code, but they support only a narrow range of domain-specific processors [Liem97, Leup97, Suda98].

Nevertheless, also machine-specific code optimization techniques should be developed with the idea of retargetability in mind. That is, new optimization techniques should be generalized as much as possible without affecting code quality.

2.7   COMPILATION SPEED

Compilers are usually expected to be very fast. For instance, the GNU C compiler on a Sun Ultra-1 workstation has a compilation speed in the order of a few thousand source lines per CPU second (including file I/O), and this speed is even comparatively low for a general-purpose system compiler.

Such compilers (e.g., used for PCs and workstations) mostly make use of optimization techniques that show a linear or low-degree polynomial runtime complexity. The result are quick compilers, that try to produce the best code possible within a quite restricted amount of time. This is reasonable, whenever large software packages need to be continuously maintained and updated, and when efficiency of the compiled code is not of major concern.

However, the situation is different for embedded systems, and therefore compilers for embedded processors should not be designed under the constraint of very high compilation speed. First of all, the efficiency required for embedded systems justifies higher compilation times. For instance, imagine a compiler that manages to reduce the code size by 50 % at the expense of spending an additional hour of optimization time for some program stored in an on-chip ROM of an embedded system product. The ROM contents will probably never be changed during the product lifetime, so that this effort will pay off in the form of lower fabrication costs for a potentially large number of devices sold. Likewise, this effort might save many man-hours of assembly programming that would be required otherwise to achieve the same level of optimization.

Secondly, the design of embedded systems in general also involves the design of hardware components. This design process implies time-intensive tasks like logic synthesis or transistor-level simulation which, even when supported by CAD tools, may take hours or days for complex hardware components. Thus, hardware design is often the actual design time bottleneck in embedded system design, and there is not really a need for very fast compilers for embedded processors. In fact, for a software-dominated embedded system, a good compiler can very well contribute more to overall efficiency than a hardware synthesis tool, and many designers are actually willing to allow for relatively long compilation times if the result is efficient code.

This trend can already be observed, for instance, in the case of the TI C6201 ANSI C compiler, which spends significantly more time in optimization and code generation than a usual compiler for a general-purpose processor. For some complex sample C program, we measured a compilation time of 120 seconds, where the GNU C compiler on the same platform finished ten times faster.

3.   OVERVIEW OF RELATED WORK

3.1   MACHINE-INDEPENDENT OPTIMIZATIONS

Almost any compiler performs a certain set of machine-independent code optimizations. Here, "machine-independent" means that optimizations are performed on an *intermediate representation* (IR) of the source program that is generated by the source language frontend at the beginning of the compilation process. This IR, typically represented as a sequence of simple three-address assignments and jump statements, includes (almost) no machine-specific information but is merely a low-level form of the original code. On this IR, one can perform a number of optimizing transformation that – most likely – will eventually result in more efficient machine code while retaining the original program semantics. Examples for such optimizations are:

Common subexpression elimination: If the IR of a source program is generated on a statement-by-statement basis, as it is usually the case, then the IR in general contains a lot of *redundant* computations. A computation is redundant at some point in a program, if that computation has already been performed at an earlier program point. Such redundancies can be eliminated by keeping the result of the first occurrence of the computation in a register and reusing the register contents instead of performing a recomputation.

Dead code elimination: Any computation that generates a result which is guaranteed to be never used again in the program can be eliminated without affecting the program semantics. Such computations are called "dead". Dead code can occur, for instance, if variables are initialized in their declaration and are later re-initialized before their first use. Also other IR optimizations generally "kill" some code, e.g., common subexpression elimination may make a lot of computations dead.

Loop-invariant code motion: Computations contained in a loop, whose results do not depend on other computations performed in that loop are loop-invariant and can thus be moved outside of the loop code in order to increase performance. Loop-invariant code in the IR is common, for instance, in loops that perform computations on array variables. Parts of the corresponding address computations are frequently independent of the actual computations performed on the array elements and can thus be moved outside the loop body.

Constant folding: Computations that are guaranteed to result in constants can be moved from program runtime to compile-time. In this case, the compiler "folds" computations on constants to single constants already when generating assembly code. Constant folding is very useful, for instance in address computations for array accesses: The index expression of an array access usually needs to be scaled by some constant in order to map the symbolic address into a physical memory address. If the index expression itself happens to be a constant, then the scaling operation can be executed already at compile time.

Many optimizations of this kind are well-known and can be found in standard compiler literature, e.g. in the classical "Dragon Book" [ASU86] or in more recent textbooks on optimizing compilers [WiMa95, Much97, Morg98, Appe98].

Even though these techniques are normally considered machine-independent, they sometimes have to be used carefully. Common subexpression elimination, for instance, reduces the number of computations to be performed, but on the other hand results in a higher number of registers required. On a processor with many functional units but only few registers, multiple recomputation of values can thus be more efficient. Likewise, constant folding may affect the program semantics when generating code for a processor whose number representation differs from that of the host machine. Constant folding then has to be restricted to those cases, where the result is guaranteed to be identical on both machines.

IR optimizations are definitely very useful in most cases in order to generate optimized machine code. However, as they do not take the detailed target machine architecture into account, additional machine-specific optimizations are also required. The machine-specific techniques presented in this book can thus be considered complementary to IR optimizations.

3.2   STANDARD CODE GENERATION TECHNIQUES

Code generation is the process of mapping machine-independent IR statements to machine-specific assembly instructions. The goal is to generate the most efficient assembly program whose functionality is equivalent to that of the source program. Due to complexity reasons, code generation is usually decomposed into several phases:

Code selection: In this phase, it is decided **which** assembly instructions will be used to implement the IR statements. Usually, code selection should not work on a statement-by-statement basis, because there is no one-to-one correspondence between IR statements and assembly instructions. The optimization goal is to select a set of assembly instruction instances whose accumulated costs, with respect to a given metric, are minimal.

Register allocation: The register allocation phase determines **where** the results of computations are stored, either in memory or in a register. More precisely, since the number of available physical registers are limited, register allocation decides which program values need to share a register such that the number of registers simultaneously required at any program point does not exceed the physical limit. Register allocation generally adds additional instructions to the previously generated code, since in some cases it might be unavoidable to *spill* and *reload* register contents to/from memory. The optimization goal is to hold as many values as possible in registers in order to avoid expensive memory accesses.

Instruction scheduling: This phase decides **when** (in terms of instruction cycle numbers) instructions will be executed at program runtime. During this, dependencies between instructions as well as limited processor resources and pipelining effects have to be taken into account. The optimization goal is to minimize the required execution time for the schedule. Instruction scheduling is particularly important in case of processors with instruction-level parallelism in order to make the most efficient use of available resources.

Like for IR optimizations, a number of standard techniques are available for the different code generation phases. We will outline one such technique for each of the three phases mentioned above in order to illustrate their benefits and limitations.

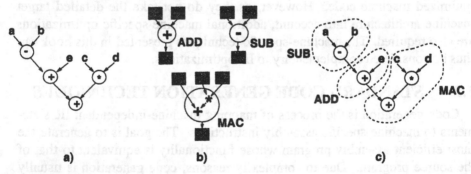

Figure 1.4. Visualization of code selection: a) data flow tree, b) instruction patterns, c) possible code selection

Tree parsing: Code selection is closely related to *tree pattern matching*. The IR of any program can be represented as a set of *data flow trees* (DFTs). A DFT represents the flow of values between computations in the following way: Each node corresponds to one computation, while each edge reflects that the result of one computation is required as an argument of another computation. Similarly, the behavior of available machine instructions can be represented by tree patterns. Using these notations, the problem of optimal code selection for a DFT is conceptually the same as the problem of *covering* the DFT by a cost-optimal set of instruction pattern instances, such that each DFT node is covered by exactly one instance. This is illustrated in fig. 1.4

The most widespread technique that implements this idea is *tree parsing* [AGT89, FHP92a]. Like in the case of string sets (or languages), one can capture the structure of *tree languages* by special grammars. Since we will make use of this approach as a subroutine, a more detailed description will be given later in chapter 3. The rules of such *tree grammars* essentially correspond to the instruction patterns. Thus, using some cost metric, one can select optimal code for a given DFT by computing the optimum *derivation* of that DFT in the grammar representing the target instruction set. Tree parsing is very efficient, since it requires only linear time in the DFT size. However, the main restriction of code selection by tree parsing is that only one DFT at a time is handled. This means that dependencies between different DFTs, as well as the interfaces of DFTs (i.e., the common subexpressions) are not well taken into account. In turn, this may result in low code quality from a global viewpoint, in particular for processors with irregular data paths.

Graph coloring: A popular approach to the register allocation problem is based on a graph model [Chai82, Brig92]. Assembly instructions generated during the code selection phase usually do not refer to concrete physical registers, but they are based on the assumption of an unlimited number of *virtual registers* for storing values. Thus, register allocation can be considered as the problem of mapping virtual registers to a fixed number of physical registers, while generating spill code in case that the physical limit cannot be met.

An instance of the register allocation problem can be represented as an *interference graph*. In this model, each node corresponds to one virtual register, while each (undirected) edge represents a *lifetime overlap* between a pair of such registers. We say that a virtual register is *live* at some program point, if its value might still be used at some later point, so that the value should be kept in a register. The lifetimes of two virtual registers overlap, if there is at least one program point, where both are simultaneously live. Thus, the interference graph represents the relation that two virtual registers must *not* be mapped to the same physical register.

If we call the physical register numbers "colors", and there are k registers available, then register allocation is the task of coloring the interference graph with at most k colors in such a way that adjacent nodes are assigned different colors. An example is given in fig. 1.5.

Unless considering only special cases, such as *interval graphs*, optimal graph coloring is an NP-complete[3] problem, so that one has to resort to

---

[3]The set NP includes all decision problems that can be solved in polynomial time by a nondeterministic Turing machine, while P is the corresponding set for deterministic Turing machines. A decision problem Π

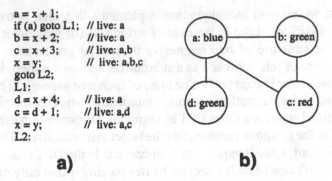

```
a = x + 1;
if (a) goto L1;    // live: a
b = x + 2;         // live: a
c = x + 3;         // live: a,b
x = y;             // live: a,b,c
goto L2;
L1:
d = x + 4;         // live: a
c = d + 1;         // live: a,d
x = y;             // live: a,c
L2:
```

a)   b)

Figure 1.5.   Register allocation by graph coloring: a) code segment annotated with live variable information, b) corresponding interference graph with a possible 3-coloring

heuristics. One heuristic makes use of the fact, that any interference graph node with a degree less than k is non-critical, since there will always be a color left for such a node. Thus, one can step-by-step remove such nodes from the graph, so as to reduce the problem size. If no further removable node exists, then one node has to be selected for spilling. Spilling essentially means reducing the live range of a node, so that further nodes may get removable. A common heuristic is to avoid spilling of nodes representing values computed in a loop, since this would induce a large number of memory accesses at program runtime. In the context of embedded processors, the most important restriction of the graph coloring approach is that it is based on the assumption of a homogeneous register set. Thus, it is not directly applicable to the special-purpose registers commonly found in DSPs.

List scheduling: Similar to register allocation, optimal instruction scheduling is an NP-hard problem in most practical cases, where resource constraints have to be obeyed. A powerful heuristic for instruction scheduling is list scheduling [DLSM81], which schedules instructions one after another. The input to instruction scheduling is a *dependency graph*, whose nodes n_i represent the instructions to be scheduled, while any directed edge (n_i, n_j) represents the fact, that n_i needs to be scheduled earlier than n_j. In a more general formulation, the dependency edges may be weighted with the *delay*

---

is *NP-complete*, if $\Pi \in NP$, and Π is "at least as difficult" as all problems in NP, which can be formally proven by constructing a polynomial-time transformation between problems. In code generation, one usually has to deal with optimization problems. Optimization problems with cost functions, that can be computed in polynomial time (which is the case for all problems relevant here), can be solved in polynomial time, if their decision counterpart can be solved in polynomial time. Optimization problems with an NP-complete decision counterpart are called *NP-hard*. For the detailed theory cf. [GaJo79]. Exactly solving NP-hard and NP-complete problems most likely (unless $P = NP$, which is a major open problem in complexity theory) requires exponential-time algorithms.

of instructions, i.e., the number of instruction cycles required to completely execute instruction n_i.

In each step, the list scheduling algorithm picks one node n_i from the current *ready set*. This set contains all nodes ready to be scheduled, i.e., nodes whose predecessors in the dependency graph have already been processed. Node n_i is then placed into the earliest possible instruction cycle c (while maintaining the current partial schedule as a list of control steps), such that

- n_i will not be executed before its graph predecessors have finished their execution
- there are sufficient free resources for n_i in cycle c.

Finally, n_i is removed from the dependency graph, and the ready set is updated. This is repeated, until all nodes have been scheduled.

A crucial issue in the effectiveness of list scheduling is the *priority function*, which selects one of possibly multiple ready nodes in each step. Common heuristics include the selection of the node with the largest number of graph successors (in order to maximize the ready set for the next step), or the selection of the node with a minimum ALAP[4] time.

List scheduling is a very effective and general instruction scheduling technique. However, its requires that all instructions to be scheduled are known in advance. Thus, it cannot directly be applied to processors, where additional instructions need to be generated *during* the scheduling phase itself. This problem will be discussed in chapter 4.

Standard code generation techniques like the ones described above, although widely used for general-purpose processors, are frequently not sufficient for embedded processors. This is mainly due to three reasons:

1. In most cases, fast **heuristics** are used for code generation. This is due to the demand for fast compilers for general-purpose processors. Since heuristics explore only a small part of the solution space of a code generation problem, the code quality may be compromised. For embedded processors, where the efficiency of the generated code is of major concern, possibly slower code generation techniques capable of exploring more solutions should be preferred.

2. The decomposition of code generation into code selection, register allocation, and instruction scheduling phases also affects code quality, since one can easily see that these **phases are mutually dependent:**

---

[4]As late as possible: The ALAP time of a node n_i denotes the maximum instruction cycle, in which n_i needs to be scheduled, such that the schedule length will not exceed the critical path length of the dependency graph. Thus, the ALAP time is a metric for the "urgency" to schedule a node.

- In case of a non-homogeneous register set, an unfavorable code selection, even though locally optimal, may induce a large number of spills during register allocation.

- Likewise, the instructions generated during code selection might not be efficiently schedulable, because code selection does not take into account the resource conflicts between potentially parallel instructions.

- Finally, instruction scheduling has a large impact on the liveness of virtual registers, so that it might be favorable to revise the register allocation on the basis of schedulability information.

Ideally, all code generation phases therefore should be executed simultaneously in order to avoid an early introduction of unnecessary restrictions in the solution space. However, such a *phase coupling* approach is both difficult to implement and time-consuming, so that this approach is rarely used in practical compilers. Techniques have been reported that partially couple register allocation and scheduling [GoHs88, BEH91, BSBC95], but so far mainly RISC processors with regular architectures have been considered.

3. Standard code generation techniques hardly take into account the **special architectural features** of embedded processors, but are frequently based on over-simplified machine models. In the following section as well as in the next chapters, we will give a number of examples for such features, which not yet handled by standard compiler technology. Additional techniques are definitely required in order to make the most efficient use of embedded processors.

3.3   CODE GENERATION FOR EMBEDDED PROCESSORS

The fact that current compilers for embedded processors tend to produce inefficient code, as well as the observation that this is largely due to missing compiler technology tuned for embedded processors triggered a number of research activities since the beginning of the 90s. As a result, special workshop series have been launched (including LCTES [LCT99], CASES [CASE99], and SCOPES [SCOP99]), and with the increasing awareness of compiler-related problems in embedded system design, many design automation conferences have listed "compilers for embedded systems" as a topic of interest. Additionally, first books focusing on code generation for embedded processors have appeared [MaGo95, Leup97, Liem97]. In the following, we will mention some important achievements. More detailed references will be given in the remaining chapters.

Rimey and Hilfinger [RiHi88], Wess [Wess91, Wess92], and Hartmann [Hart92] were among the first to present code generation techniques targeted towards irregular processor architectures. The proposed techniques are capable of generating efficient code for processors with special-purpose registers and instruction-level parallelism. A key idea is to minimize the number of register-to-register move and spill instructions for special-purpose registers. A more formal approach, based on parallel language parsing has been described in [MME90], but it has only been applied to simplified processor architectures. The same holds for the automata-model based approach from [LaCe93]. Kolson et al. have described a special technique for allocation of special-purpose registers in loops [KNDK95]. A complete compiler for irregular processor architectures has been developed within the CodeSyn project [LMP94a, LMP94b, LPCJ95]. However, many ad hoc techniques have been used, and the applicability is restricted to very special embedded processor classes.

A different approach has been proposed by Philips researchers [SMT+95, TSMJ95]. Here, the main goal is efficient instruction scheduling for DSPs based on an extensive pre-analysis of available scheduling options. In [LeMa95a], a related scheduling technique based on *Integer Linear Programming* has been presented, capable of exactly meeting time constraints for DSP algorithms. This technique has also been applied to optimized code selection and scheduling in presence of strongly encoded instruction formats [LeMa96a].

Several important contributions to code generation for embedded processors came from the SPAM project. These include an adaptation of the tree parsing technique for code selection to irregular architectures [ArMa95], minimization of *mode switching* instructions in DSPs [LDK+95a], and partitioning of program variables among multiple memory banks [SuMa95].

Also the problem of phase coupling has been considered in different approaches. In the CHESS compiler, *data routing* [LCGD94] has been used as a technique to simultaneously solve the problems of code selection and register allocation, while in the AVIV compiler [HaDe98], code selection has been coupled with scheduling and the assignment of values to distributed register files. A technique aiming at a complete coupling of code generation phases is Mutation Scheduling [NND95], which also incorporates the use of *algebraic transformations* in order to generate efficient code. The main difficulty, however, is the efficient control of the search space exploration. The Integer Linear Programming based formulation in [WGHB94] also is a completely phase-coupled technique. However, it suffers from extremely high runtime requirements.

A very different approach has been taken in [BaLe99a], where *constraint logic programming* has been exploited in order to implement phase coupling. The main idea is to model the constraints on code generation (i.e., the things the compiler must *not* do) that are imposed by the processor architecture, and to generate code by labeling a set of solution variables such that all constraints are met. By using appropriate cost metrics, an optimized labeling and thus optimized code can be generated. In [MPE99], a constraint-based technique for coupling register allocation and scheduling under real-time constraints has been described. There, constraint analysis is applied in order to avoid register spills.

Rau et al. have presented a fine-grained approach to phase coupling, specifically designed for a class of VLIW processors [RKA99]. Chen and Lin [ChLi99] proposed a fast algorithm for phase-coupled scheduling and register allocation for a specific DSP, which achieved a significant code performance increase as compared to a commercial C compiler.

Another area that has received significant attention is optimized *memory address computation*. This problem arises due to the special address generation units in DSPs, which are capable of computing memory addresses in parallel to the central data path in certain situations. Important contributions to this area include [Bart92, LDK+95b, LeMa96b, ASM96, LPJ96, WeGo97a]. More detailed descriptions will be given in chapter 2, which specifically deals with memory address computation.

3.4   RETARGETABLE COMPILERS

Although there exist a number of general-purpose retargetable compilers, such as the GNU compiler gcc [GNU00] and lcc [FrHa95], these can hardly be applied to code generation for embedded processors, since they have been designed primarily for a class of general-purpose 32-bit processors. Therefore, retargeting to domain-specific architectures is very difficult and the generated compilers have problems with special-purpose registers, instruction-level parallelism, and the memory access structure. This is also confirmed by the DSPStone project [ZVSM94], where several GNU-based C compilers for DSPs turned out to generate code of very low quality.

Since retargetable compilers are very important in the context of ASIP-based designs (section 1.5), several research projects aimed at providing retargetable compiler technology for embedded processors. The MSSQ compiler [Nowa87, Marw93, LeMa98] is capable of generating microcode for almost arbitrary data path architectures modeled in an HDL. In this way, a close link to processor hardware design is ensured, because HDL models can also be used for synthesis and simulation purposes. By using a special "bootstrapping" technique [LSM94], the capabilities of MSSQ have been extended beyond pure microcode generation. However, as a result of its high degree of retargetability, the code quality of MSSQ is rather low. This has been changed in the RECORD compiler [Leup97], which uses HDL models like MSSQ, but focuses on DSP architectures only. Therefore, it has been possible to implement code generation techniques, which are simultaneously retargetable and efficient for a class of processors. In order to capture a wider range of HDL models, RECORD uses an *instruction set extraction* technique [LeMa95b] in order to determine the target instruction set in advance. From the extracted instruction set, a target-specific code generator is automatically generated [LeMa97].

A different approach has been taken in the CBC [FaKn93] and CHESS [LVK+95] compiler projects, where the special processor modeling language nML [FVM95] has been used to achieve retargetability. Architectures captured by nML are primarily load-store DSP architectures. The code quality is satisfactory, however at the expense of only a small range of supported processor architectures. Similarly, the AVIV compiler [HaDe98] uses ISDL [HHD97] as a special language for describing VLIW-like embedded processors. Likewise, the compiler described in [RKA99] is retargetable within a class of VLIW processors based on a special machine description formalism.

Also the CodeSyn [Liem97] and SPAM [Suda98] compilers are retargetable to a certain extent. However, the main goal in these projects was high code quality. Therefore, target processors are not described in the form of unified models, but in a very heterogeneous way, partially including source code for target-specific optimization routines.

There are also several commercial tools capable of retargetable compilation for embedded processors: The CHESS compiler [LVK+95], together with a retargetable assembler and an instruction set simulator has been commercialized by Target Compiler Technologies [Targ00]. However, also this version supports only a narrow class of target architectures. ACE [ACE00] provides the CoSy system, which can be regarded as a compiler construction toolbox. The system comprises language frontends and standard optimizations, while target-specific code generators and optimizers can be generated and "plugged in" as long as they are compatible with the common IR in CoSy. Archelon [Arch00] provides a truly retargetable C compiler that directly works on a textual description of the target processor. However, code generation is based on standard techniques only, and it is relatively difficult to bypass these built-in techniques in order to perform machine-specific optimizations.

4. STRUCTURE OF THIS BOOK

The goal of the work presented in the following chapters is to increase the quality of compiled code for embedded processors by novel code optimization techniques. The motivation is that only such techniques, which take the special constraints in embedded system design into account, will permit to replace time-consuming assembly programming of embedded processors by the use of compilers. Eventually, this will enable a significantly higher productivity in the design of embedded systems based on programmable processors.

We will mainly consider two classes of embedded processors, DSPs and multimedia processors, both for which a more advanced compiler technology is definitely required. Chapters 2 and 3 deal with DSPs, while chapters 5 to 7 are targeted towards multimedia processors.

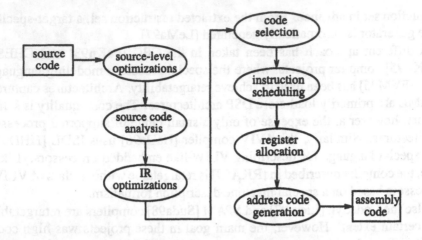

Figure 1.6.   Coarse compilation flow

The logical structure of this book, in a bottom-up fashion, also follows the different levels in the compilation flow at which code optimizations can take place. The organization of compilation phases is outlined in fig. 1.6. This is only a coarse view of the compilation flow, as in practice the phase ordering may be partially changed and phases may also be coupled.

For DSPs, which tend to show a quite irregular architecture, low-level optimizations are required that take the detailed hardware into account in order to generate efficient code. For multimedia processors, whose architecture is generally more RISC-like (and thus more regular than in DSPs), classical low-level code generation techniques (such as register allocation by graph coloring) perform well and leave little room for improvements. For those processors, we will therefore primarily present optimization techniques operating at higher levels of abstraction, which are targeted towards processor features not yet sufficiently covered by standard compiler technology. From chapter to chapter, the abstraction level will be increased step-by-step, until in chapter 7 we present a pure source-level optimization.

Our main optimization goals are high performance and/or low size of compiled code. For this purpose, we will frequently utilize comparatively time-intensive basic optimization techniques, which are rather unconventional in classical compiler construction. These include genetic algorithms, simulated annealing, branch-and-bound, and integer linear programming. Using such powerful optimization techniques allows us to solve complex code optimization problems for which good heuristics are out of sight. Still, one has to take care of the efficiency of the compilation process. Therefore, whenever possible, we will rely on subroutines implemented by efficient standard techniques, problem-specific heuristics, and fast special-purpose algorithms.

As we will show experimentally in each chapter, this approach achieves significant improvements in code quality within compilation times that are still acceptable in embedded system design. Since we aim at providing compiler techniques that work in practice, we will give experimental results for real-life embedded processors.

We will now provide a brief overview of the contents of the following chapters.

- **Chapter 2** presents techniques for optimized **memory address computation** for DSPs. This chapter is divided into two parts. The first part deals with different variants of the **offset assignment problem** for scalar variables, for which we present a unified optimization technique. The second part describes a new technique for **address register allocation** for array accesses in loops.

- **Chapter 3** deals with the problem of **register allocation in DSP data paths**. We will focus on the problem of allocating special-purpose registers for **common subexpressions** in data flow graphs in order to increase code efficiency with respect to several quality metrics.

- **Chapter 4** deals with multimedia processors with a VLIW-like data path divided into **functional clusters** with limited connectivity. This clustering, which is required for efficiency reasons, has strong implications on **instruction scheduling**. We will present a scheduling technique that improves code performance by incorporating the task of **partitioning** instructions between the clusters into the scheduling process.

- **Chapter 5** treats a new class of instructions found in recent multimedia processors, that resemble the SIMD paradigm in computer architecture. These **SIMD instructions** allow for an efficient utilization of functional units for parallel computations on certain data types. However, they cannot be exploited in current compilers without the need for assembly libraries or "compiler intrinsics". In contrast, the presented **code selection technique** allows to take advantage of SIMD instructions also for plain C source code.

- **Chapter 6** presents a performance optimization for **control-dominated** applications, i.e., applications whose source code comprises many if-then-else constructs. Recent embedded processors offer hardware support for if-then-else statements by means of **conditional instructions**. These instructions allow to eliminate conditional jumps and thereby reduce the number of instruction pipeline stalls. However, conditional instructions are not always better than conditional jumps. We will analyze the trade-off between these two alternatives, and we will present a dynamic programming algorithm that systematically generates performance-optimized code for nested if-then-else statements.

- **Chapter 7** describes a technique that reuses the well-known concept of **function inlining** for performance optimization at the source code level. Unlike in existing inlining techniques, mainly based on heuristics, our approach aims at a maximum program speedup by inlining under a **global code size limit**. Thus, it is useful for embedded processors with a limited amount of program memory.

- **Chapter 8** discusses some **compiler frontend aspects** from a practical point of view. Many of the optimization techniques presented here have been evaluated with help of the **compiler development platform LANCE**. LANCE mainly consists of an ANSI C frontend and a library of machine-independent standard code optimizations. The system turned out to be very useful in practice and has also been successfully applied for academic and industrial compiler projects beyond this book. We will therefore briefly outline the software architecture and the features offered by the LANCE system.

Chapter 2

MEMORY ADDRESS COMPUTATION FOR DSPS

In this chapter, we describe two code optimization techniques for DSPs, which exploit the special address generation hardware in such processors. These optimizations are very "low-level" in the compilation flow, because they demand that machine code, in the form of sequential assembly code with symbolic variables, has already been generated in earlier phases (fig. 2.1). One degree of freedom that can be exploited at this point is the way in which symbolic variables are laid out in memory and how the memory addresses of the variables are computed at program runtime.

After describing the address generation hardware typically found in DSPs, we will present a novel algorithm for solving the *offset assignment problem*, which aims at an optimum placement of scalar variables in memory. In contrast, the last section deals with the problem of optimized *address register allocation for array accesses* in loops.

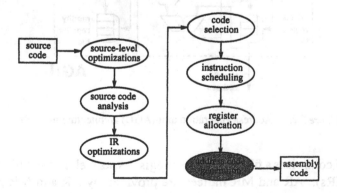

Figure 2.1. Address code generation in the compilation flow

1.   ADDRESS GENERATION HARDWARE

In chapter 1, we have outlined some characteristics of DSPs. These processors are tailored towards fast execution of arithmetic-intensive signal processing algorithms. On the other hand, for efficiency reasons, DSPs offer lower programming comfort than other processor families. In particular, this concerns the variety of available *addressing modes*. In general, DSPs offer mainly two such modes: direct and indirect. The direct addressing mode uses an immediate field in the instruction word to form memory addresses, while in indirect mode, addresses are read from *address registers* (ARs). More precisely, a pointer stored in one certain AR, without any further modification, has to be used for memory addressing. More complex modes, such as "base plus offset" addressing which are common for CISC architectures, are mostly not provided. The reason is, that such modes would significantly contribute to the execution time of an instruction (either via additional instruction cycles or via a longer combination delay in the data path), which is not desirable for DSPs.

Instead, DSPs permit to compute certain addresses *in parallel* to the central data path. This means that a memory address required in instruction cycle n sometimes can be computed by an AR modification in cycle $n - 1$ (or earlier) without the need for extra instructions. Such parallel address computations are supported by a dedicated *address generation unit* (AGU), as shown in fig. 2.2.

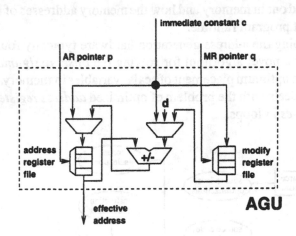

Figure 2.2.   Address generation unit (AGU) architecture in DSPs

The AGU comprises a file of k address registers, as well as a file of m *modify registers* (MRs). AR and MR indices are provided by AR and MR *pointers*, which are either part of the instruction word or are stored in special registers. For a given AR pointer value p, AR$[p]$ is the "current" AR holding the effective address for memory access in indirect mode. According to the register-transfer structure shown in fig. 2.2, there are the following ways of computing addresses by modifying an AR in the AGU:

immediate load: A constant c is loaded from an immediate instruction field into the current AR.

immediate modify: A constant c from an immediate instruction field is added to (or subtracted from) the current AR.

auto-increment: The current AR is incremented (or decremented[1]) by a constant d. The absolute value of d is bounded by some "small" constant r.

auto-modify: The current AR is incremented (or decremented) by the contents of the current modify register $MR[q]$ for some value q of the MR pointer.

Like ARs, also MR contents can be changed by loading an immediate constant c into the current MR. MRs serve as containers for a set of constants frequently required for AR modification.

The only address computations *not* requiring an immediate constant are auto-increment and auto-modify. These two are also the only address computations that can be executed in parallel to other machine instructions. The reason is that encoding of the short immediate constant d does not require an extra instruction word. If b bits are available for encoding d, then d is bounded by $r = 2^{b-1}$ (or $2^b - 1$ in case of unsigned modifiers). Auto-modify operations do not require any immediate constant at all, since an MR can be encoded implicitly.

All other address computations consume an extra machine instruction and thus increase code size and reduce performance. For instance, an immediate modify is required, whenever the modifier c is larger than r, so that c has to be encoded in a separate word. Therefore, the goal in address code generation for DSPs must be to *maximize the use of auto-increment and auto-modify operations* for generating addresses.

The AGU architecture shown in fig. 2.2 is almost identical for many important DSP families (such as Texas Instruments C2x/C5x, Motorola 56xxx, and Analog Devices ADSP-210x), as well as for a number of ASIPs (e.g. the Gepard core mentioned in chapter 1). The main differences lie in the values of the *AGU parameters k* (number of ARs), *m* (number of MRs), and *r* (auto-increment range). Thus, retargetability of address code optimization can be achieved by designing algorithms that take k, m, and/or r as parameters instead of using fixed values. Table 2.1 show the concrete AGU parameters of some popular DSPs.

---

[1]For sake of simplicity, we will only use the term "auto-increment" to denote an address modification by ± 1, unless we explicitly need to distinguish between increment and decrement.

Table 2.1.   AGU parameters of different DSPs

	TI C25	Motorola 56xxx	ADSP-210x	AMS Gepard
k	8	4	4	8
m	1	4	4	8
r	1	1	0	7

2.   OFFSET ASSIGNMENT
2.1   BASIC PROBLEM DEFINITION

The restricted number of addressing modes in DSPs are frequently sufficient for assembly-level programming of DSP algorithms, but they have significant consequences for the design of C compilers. As in most programming languages, C has the concept of *functions* and *local variables*. These variables exist only during the execution of some particular function at program runtime and therefore have to reside in a runtime stack or, if possible, in registers. For DSPs, which have a very limited amount of registers, it is reasonable to assume that in fact most local variables have to reside in the stack.

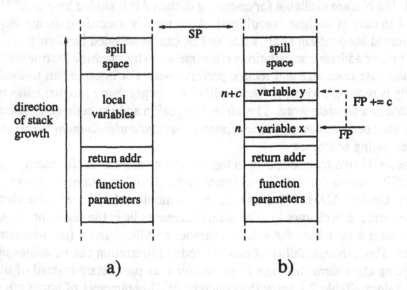

Figure 2.3.   a) Stack layout during function execution: The stack contains the function parameters (pushed by the calling function), the return address (pushed by the call instruction), the local variables, and a spill space for temporarily saving register contents. The stack pointer points to the next free memory cell in order to allow for further function calls. b) "Floating" frame pointer: FP must be incremented or decremented by an offset difference c for each variable access.

Fig. 2.3 a) shows a typical stack layout when some function is being executed. As the stack changes its structure and size dynamically, all data on the stack have to be addressed relative to the stack pointer (SP). Thus, an "SP plus offset" addressing mode would be useful, but is not available on DSPs with an AGU as in fig. 2.2. The solution to this problem is to use an additional *frame pointer* (FP), which dynamically moves through the stack frame of the function[2]. In the beginning, FP is initialized with the effective address of the first local variable being accessed in the function. Then, for each subsequent access, FP is incremented or decremented by the corresponding offset difference (fig. 2.3 b). Naturally it is useful to keep FP in an address register.

The stack position of function parameters and the return address are normally fixed (since the function might be called from a external C module, and the return address has to be present at a certain location when returning from the function), but the compiler has the freedom to permute the local variables (and also the spill locations) in the stack frame. This can be exploited by arranging the local variables in such a way, that the required FP updates can be implemented by auto-increment AGU operations in many (or even most) cases. The corresponding *offset assignment problem* can be formulated as follows: Given a set of local variables

$$V = \{v_1, \ldots, v_n\}$$

and the access sequence

$$S = (s_1, \ldots, s_l), \quad \text{with} \quad \forall i \in \{1, \ldots, l\} : s_i \in V$$

to the variables in V, find a bijective *offset mapping*

$$M : V \to \{0, \ldots, n-1\}$$

such that, for a given auto-increment range d, the *cost function*

$$C(M) = 1 + \sum_{i=1}^{l-1} z_i$$

is minimized, where

$$z_i = \begin{cases} 1, & \text{if } |M(s_{i+1}) - M(s_i)| > r \\ 0, & \text{else} \end{cases}$$

The "1" in $C(M)$ is needed to count the instruction required for AR initialization, which always requires an immediate load operation. Fig. 2.4 illustrates the offset assignment problem for an example with

$$r = 1, \quad V = \{a, b, c, d\}, \quad S = (b, d, a, c, d, a, c, b, a, d, a, c, d)$$

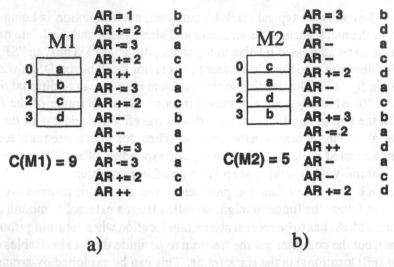

Figure 2.4. Offset assignment example: a) Suboptimal memory layout $M1$, where variables are assigned to offsets in alphabetical order. The sequence of AGU operations needed to compute the memory addresses corresponding to the access sequence S is given in C-like notation. Only 4 out of $|S| = 13$ operations are auto-increment/decrement, resulting in a cost value of $C(M1) = 9$. b) Improved layout $M2$ with $C(M2) = 5$. The cost reduction is due to the adaptation of $M2$ to S, so that most AGU operations are auto-increment/decrement.

Essentially, offset assignment is the problem of finding good memory layouts as the one shown in fig. 2.4 b). Note that the above formulation captures only a simple special case: offset assignment for one AR and zero MRs. We will later generalize this formulation towards more relevant sets of AGU parameters.

2.2   RELATED WORK

The first offset assignment algorithm was presented by Bartley [Bart92]. We will use the notation (k, m, r)-OA to denote concrete AGU parameter configurations. Bartley considered (1,0,1)-OA and modeled the problem by means of an undirected edge-weighted *access graph* $G = (V, E, w)$, where V is the set of variables, and for each edge $e = (v_1, v_2) \in E$, the weight $w(e)$ is equal to the number of transitions (v_1, v_2) or (v_2, v_1) in the access sequence S. Fig. 2.5 a) shows the access graph for our example from fig. 2.4. Bartley's algorithm is based on two observations:

1. Variable pairs (v_1, v_2) that have a high number of transitions in S should be placed into neighboring memory locations, because in this case all transitions from v_1 to v_2 and vice versa can be implemented by auto-increment or auto-decrement operations.

---

[2] This is sometimes called a *roving* or *floating frame pointer*.

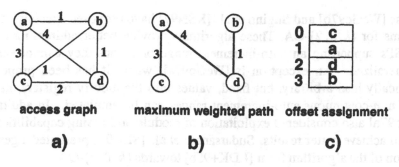

access graph   **maximum weighted path** **offset assignment**

a)   **b)**   **c)**

Figure 2.5.   Access graph model for offset assignment

2. Each concrete offset assignment corresponds to a *Hamiltonian path* in G, i.e. a path that touches all nodes once.

From these observations it is obvious that an optimum offset assignment corresponds to a maximum weighted Hamiltonian path in G (fig. 2.5 b and c). Bartley presented an $\mathcal{O}(n^3)$ algorithm that heuristically finds good paths.

In [LDK+95b], Liao showed that the offset assignment problem is NP-hard even for the simple case $(1, 0, 1)$-OA, and he presented a more efficient heuristic also based on the access graph model. He modified Kruskal's *minimum spanning tree* algorithm [HoSa87] to construct Hamiltonian paths instead of trees with a complexity of $\mathcal{O}(n \cdot \log n)$. Liao also presented a generalization towards $(k, 0, 1)$-OA i.e., an arbitrary number of ARs. With respect to the stack frame model from fig. 2.3, this means that multiple "floating" frame pointers are used simultaneously in order to reduce the offset assignment costs.

For k ARs, the variable set V has additionally to be partitioned into disjoint subsets $V_1, \ldots, V_k$, such that one AR is used for each subset V_i. In this way, $(k, 0, 1)$-OA is reduced to k problems of type $(1, 0, 1)$. However, Liao only presented a relatively simple variable partitioning heuristic.

In [LeMa96b], Liao's $(1,0,1)$-OA algorithm has been improved by a *tie-break heuristic*. The original algorithm in each step selects the next edge to be included in the Hamiltonian path. In case of equally weighted edges, the choice is normally made arbitrarily. The tie-break heuristic selects one from those edges based on a certain cost metric in order to improve the results. Additionally, [LeMa96b] described a better variable partitioning heuristic for $(k, 0, 1)$-OA, and a generalization towards $(k, m, 1)$-OA has been presented. The latter is based on the observation that a *page replacement algorithm* from operating system theory [Bela66] can be adapted to optimally exploit m modify registers. This will be described in more detail in the next section. Essentially, MRs are used to hold frequently required address modifiers larger than the auto-increment range.

Wess [WeGo97b] and Sugino et al. [KSN97] have proposed specialized algorithms for $(k, 0, 2)$-OA. These algorithms provide better addressing code for DSPs supporting an auto-increment range of 2, but they cannot easily be generalized. An exception is [WeGo97a], where it has been proposed to statically load arbitrary, but fixed, values into the modify registers, using which non-continuous auto-increment ranges can be emulated. In addition, [WeGo97a] also considered exploitation of modulo addressing capabilities in order to achieve better results. Sudarsanam et al. [SLD97] presented a generalization of the algorithm from [LDK+95b] towards $(k, 0, r)$-OA.

Several researchers have considered special extensions of the offset assignment problem. Normally, an auto-increment AGU operation can only take place in parallel to a memory access. Sugino et al. [SMIN96, SMN97] have considered machines without this restriction. In this case, an AR modification by a constant larger than r can also be implemented by multiple auto-increments placed in instruction sequences without memory access. In [ChLi98], DSPs have been considered where r is not constant but depends on the concrete AR being used. Finally, in [RaPa99], the scheduling freedom that may be present in the memory access sequence (e.g. due to commutativity of arithmetic operations) has been exploited to generate better $(1, 0, 1)$-OA solutions than in the case of a totally ordered access sequence.

Also more global extensions of the offset assignment problem have been considered. The problem of partitioning variables between multiple parallel memory banks has been considered in [PLN92, SuMa95, SCL96], while in [EcKr99] a technique for minimization of frame pointer reloads across basic block boundaries has been presented.

2.3   ALGORITHM FOR (K, M, R) OFFSET ASSIGNMENT

PROBLEM DEFINITION

As shown in the previous section, existing offset assignment algorithms show limitations w.r.t. the AGU parameter configurations that can be processed. Table 2.2 summarizes the configurations handled by different techniques.

In the following we will present an algorithm that solves the general (k, m, r) offset assignment problem, which we formulate as follows. Given a variable set V and an access sequence S with

$$V = \{v_1, \ldots, v_n\}, \quad S = (s_1, \ldots, s_l)$$

find a partitioning

$$V = V_1 \cup \ldots \cup V_k$$

Table 2.2.   AGU parameters for different offset assignment techniques

reference	# address registers	# modify registers	auto-increment range
[Bart92]	1	0	1
[LDK+95b]	k	0	1
[LeMa96b]	k	m	1
[WeGo97b]	1	0	2
[SLD97]	k	0	r
here	k	m	r

of V into k disjoint subsets as well as k offset mappings

$$M_1 : V_1 \rightarrow \{0, \ldots, |V_1| - 1\}$$
$$\cdots$$
$$M_k : V_k \rightarrow \{0, \ldots, |V_k| - 1\}$$

and a mapping

$$X : \{1, \ldots, l\} \rightarrow \mathbb{N}^m$$

such that the cost function

$$C = k + \sum_{i=1}^{l-1} z_i$$

is minimized. Let $P(s_i)$ denote the index of the variable subset which includes s_i. Then, z_i is defined by

$$z_i = \begin{cases} 0, & \text{if} \quad P(s_{i+1}) \neq P(s_i) \\ 0, & \text{if} \quad P(s_{i+1}) = P(s_i) \quad \text{and} \quad |M_{P(s_i)}(s_{i+1}) - M_{P(s_i)}(s_i)| \leq r \\ 0, & \text{if} \quad P(s_{i+1}) = P(s_i) \quad \text{and} \quad |M_{P(s_i)}(s_{i+1}) - M_{P(s_i)}(s_i)| \in X(i+1) \\ 1, & \text{else} \end{cases}$$

In this formulation, $X(i)$ denotes the set of address modifiers stored in the m MRs at the point of time when access s_i takes place. The definition of z_i reflects that there are three possibilities of exploiting auto-increment or auto-modify operations, thus avoiding costs due to extra instructions:

1. If s_{i+1} and s_i are in different variables subsets, then only the AR used for addressing needs to be switched. This means that the AR pointer needs to be reloaded. Since the AR pointer is part of the instruction word, this does not incur an extra instruction.

2. If s_{i+1} and s_i are in the same variables subset, then, like in the simple formulation given above, auto-increment can be exploited if the offset distance is less or equal to r.

3. If s_{i+1} and s_i are in the same variables subset, and the offset distance d is larger than r, then an auto-modify operation can be used if the required modifier d is currently present in some MR.

In all other cases, one extra instruction is required. This could be either an immediate modify AGU operation or an immediate load of an MR. If the latter is used, then the required address computation can be implemented by auto-modify. In addition, the loaded value might be reused for further variables accesses. However, the limited number of MRs must be taken into account.

$V = \{v0, v1, v2, v3, v4\}$

$S = (v0, v4, v1, v3, v2, v0, v2, v0, v3, v0,$
$\qquad v1, v2, v2, v2, v4, v0, v4, v0, v0, v3)$

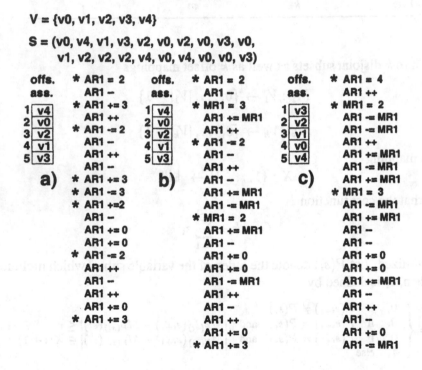

Figure 2.6. Offset assignment with MRs: a) Heuristically generated $(1, 0, 1)$-OA (address computations not covered by auto-increment are marked with "*") with a cost value of 8. b) $(1, 1, 1)$-OA based on the same layout as in a), where modify register MR1 is used to hold multiply required modify values (value 3 is used three times, then value 2 is used two times) with a cost of 5. c) $(1, 1, 1)$-OA, where the layout has been constructed in such a way that the utilization of MR1 is optimal. Only three AGU operations remain that are not covered by auto-increment or auto-modify. In fact, one can show that this is an optimal solution.

GENETIC ALGORITHM FORMULATION

An algorithm for the generalized (k, m, r)-OA formulation offers two main advantages. First, it can be applied to a large variety of AGUs, which makes it attractive for retargetable compilers. Second, the generalized formulation allows for better solutions which cannot be achieved by previous methods. This is exemplified in fig. 2.6 for $(1, 1, 1)$-OA. The key improvement is that MRs are exploited *during* offset assignment instead of allocating MRs only in a postpass phase as in previous work [LeMa96b]. Thus, it is a simple example of phase coupling.

Since (k, m, r)-OA is a very complex optimization problem (even $(1,0,1)$-OA is NP-hard), we use a *genetic algorithm* (GA) to obtain close-to-optimal solutions. GAs are especially well-suited for complex nonlinear optimization problems, because they are capable of skipping local optima in the objective function. An overview is given in [Davi91]. Among numerous other applications, GAs have been successfully used in embedded system design in the context of behavioral synthesis [LaMa97]. Drechsler [Drec98] applied GAs to a number of different VLSI design problems and concluded that problem-specific knowledge should be exploited whenever possible, so as to ensure a good performance. This idea is also used in our approach.

The main concept of GAs is to simulate natural evolution on a *population* (a set) of *individuals* (each representing a possible solution). The simulation takes place by iterating over several *generations*, in each of which the population is modified. The goal is to improve the quality of its individuals according to some cost metric called the *fitness function*. The iteration usually terminates after a fixed number of generations or if the population appears to be stable w.r.t. the fitness. GAs cannot guarantee to find optimal solutions, but it is known that solutions close to the optimum can be found when spending sufficient runtime.

Each individual is represented by a *chromosome*, i.e. a string of *genes* (symbols) that encode the properties of a solution. In fact, a secondary motivation for using a GA for offset assignment is that this problem offers a relatively straightforward chromosomal representation of solutions: For a variable set $V = \{v_1, \ldots, v_n\}$ we specify that for a chromosome $C = (g_0, \ldots, g_{n-1})$ the gene g_i has the value j if variable v_j is mapped to offset i (fig. 2.7 a). Thus, each chromosome can be regarded as a permutation of V.

This representation works for $k = 1$. For arbitrary k values we need to insert $k - 1$ *separator symbols* into the chromosome in order to assign certain variable subsets to certain ARs (fig. 2.7 b). For this purpose we can still use permutations by extending the index set to $\{1, \ldots, n, n+1, \ldots, n+k-1\}$. As above, index values less or equal n are interpreted as variable indices, but values larger than n are interpreted as a switch to the next AR. This is convenient since we can generate new valid solutions (or individuals) simply by generating new permutations.

GAs use two operators, *crossover* and *mutation*, to generate new individuals. In each generation, crossover is applied to pairs of individuals to generate an *offspring*. The idea is that in this way the "good" properties (in our case: good partial memory layouts) of both parents can be combined by inheriting parts

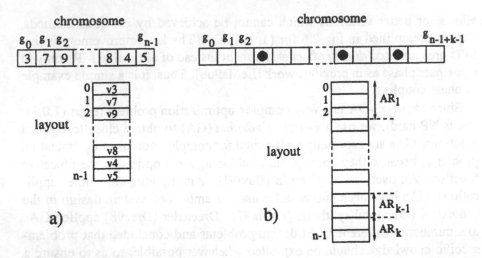

Figure 2.7. Chromosomal representation: a) for 1 AR, b) for k ARs, black circle denotes a separator

of their chromosomes to the offspring. In our algorithm the crossover must be permutation preserving. Therefore we use a standard crossover operator called *order crossover* (fig. 2.8), which has this property and works as follows:

1. Randomly choose two gene indices in the parents' chromosomes A and B. These indices induce a three-partitioning $A = (A_1, A_2, A_3)$ and $B = (B_1, B_2, B_3)$.

2. Mark the position of indices occurring in A_2 in B and the position of indices occurring in B_2 in A.

3. Generate a new chromosome A' from A: Starting with the leftmost position of interval A_2, in left-to-right order, write the non-marked indices of A into A_1 and A_3, while leaving interval A_2 as a gap. Generate a new chromosome B' from B analogously.

4. Copy the center interval from B to A', and copy the center interval from A to B'.

The mutation operator simulates the mutations that take place in natural evolution by performing random modifications of the newly generated offspring individuals. A mutation should allow to leave local optima but should also retain most of the structure of an individual. Therefore, the corresponding modifications must not be too global. In our case we use a *transposition* (fig. 2.9) as a mutation operator:

1. Randomly choose two gene indices i and j in the individual to be mutated.

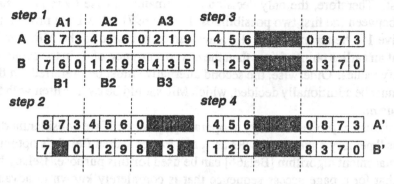

Figure 2.8. Order crossover

2. Exchange the contents of genes g_i and g_j.

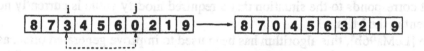

Figure 2.9. Mutation operator

The fitness function in our GA inversely reflects the cost function given at the beginning of this section: An individual with a cost value C has a fitness of $-C$, so that low cost corresponds to high fitness. In order to count those positions in the access sequence S, where auto-increment operations cannot be used, only a single traversal of S is required for a given offset assignment. However, the exploitation of MRs requires additional effort. As described in [LeMa96b], we can use a variant of a *page replacement algorithm* in order to optimally exploit m MRs for a given offset assignment, which we briefly summarize here.

Any offset assignment is characterized by a sequence of *modify values* $(d_1, \ldots, d_{m-1})$. A value d_i denotes that an AR needs to be incremented (or decremented) by d_i in order to compute the memory address required for variable access s_i. Any d_i less or equal to r can be implemented by auto-increment and thus does not incur costs. For $d_i > r$, there are three implementation possibilities:

1. Immediate AR modify operation.

2. Immediate load of an MR, followed by an auto-modify operation.

3. Auto-modify operation, if some MR currently contains d_i.

Whenever there is some MR currently containing d_i (case 3), then reusing this value is obviously always favorable, since an auto-modify operation has zero cost. Therefore, the only decision that remains in case there is no such MR, is between the first two possibilities. In [Leup97] it has been proven that alternative 1 is better exactly if all values currently stored in the MRs can be reused at an earlier point of time than the next occurrence of d_i in the sequence of modify values. Otherwise, the second alternative must be preferred. In this case it must be additionally decided, which MR should be overwritten with the new value d_i.

Now we can exploit an analogy to operating systems: If we consider modify values as "memory pages" and MRs as "page frames", then Belady's optimum page replacement algorithm [Bela66] can be used for this purpose. Belady has shown that for a page access sequence that is completely known in advance (which usually does not hold in operating systems, but fortunately in our case since the variable layout and the modify values are known), the frame containing the page with the largest distance to the next access in the future must be overwritten in order to minimize the number of page faults. In our case, a page fault corresponds to the situation that a required modify value is currently not present in the MRs.

In [LeMa96b], this algorithm has been used to improve generated offset assignments in a postpass phase. The possible improvement has been exemplified in fig. 2.6 b), but the benefit is limited if the offset is already fixed.

In the GA, however, we can directly incorporate the exploitation of MRs *into the fitness function*. As a consequence, the generated offset assignments are not only tailored towards exploitation of auto-increment but simultaneously take the available MRs into account. This gives much better optimization opportunities and, as exemplified in fig. 2.6 c), may even result in completely different memory layouts. With a careful implementation, the algorithm for MR exploitation takes only linear time in the access sequence length and thus does not increase the asymptotic complexity of the fitness function.

Fig. 2.10 gives the global GA procedure for (k, m, r)-OA in a pseudo-code notation. First, an initial population is randomly generated. In order to accelerate the convergence of the GA by a "seed", one initial individual is generated with a heuristic [LeMa96b]. The fitness of the initial population is computed according to the above cost function, and then G_1 generations are simulated. In each generation, a set of parent individuals are "married"[3] to generate an offspring using the order crossover operator. Then, a fraction of the offspring is mutated and the fitness is evaluated again. Finally, for some parameter p, a percentage p of the current population with lowest fitness are replaced by the best offspring individuals, and the next generation is simulated. An additional termination condition for the loop is that the last G_2 (with $G_2 < G_1$) gener-

---

[3] We use the *tournament selection* scheme [Davi91] for this purpose.

```
algorithm OFFSETASSIGNMENT
input: variable set V, access sequence S, AGU parameters k, m, r
output: offset assignment
begin
    GENERATEINITIALPOPULATION();
    EVALUATEFITNESS();
    for G₁ generations do
        SELECTPARENTS();
        GENERATEOFFSPRING();
        MUTATEOFFSPRING();
        EVALUATEFITNESS();
        REPLACEPOPULATION();
        if no max fitness improvement in the last G₂ generations then break // exit loop
        end if
    end for
    return best individual;
end algorithm
```

Figure 2.10.   Genetic algorithm for (k, m, r)-OA

ations did not result in an improvement of the maximum fitness. In this case, the population is considered stable (i.e., either the optimum has been found, or the GA is trapped in a local minimum from which it will probably not escape anymore), and the loop is exited.

2.4   EXPERIMENTAL RESULTS

The GA for (k, m, r)-OA has been evaluated on a large number of random problem instances. An experimental evaluation using random inputs bears the disadvantage that we do not get results for "real" problems. However, we prefer this method here for two reasons:

1. Showing that the technique produces good results on the average indicates that it will generally also achieve good results for "real" problem instances.

2. Using a sufficiently large input data base ensures the reproducibility of results, even without having access to the detailed benchmarks. This strongly facilitates the experimental comparison of (possible) new improvements against previous techniques. In contrast, "real" benchmarks used in the literature are frequently not publicly available, or they exist in various versions.

Experiments [Davi98] showed that the GA parameter values from table 2.3 are a good compromise for different AGU configurations and problem instances:

Table 2.3.   GA parameters for experimental evaluation

parameter	value
Population size	30
Max number of generations G_1	5000
Second termination condition G_2	2000
Mutation probability per offspring individual	$1/(n + k - 1)$
Replacement rate p	2/3

In practice, this results in a typical runtime of about 10 CPU seconds on a workstation[4], which can be regarded as acceptable, since for embedded processors we do not require high compilation speed.

The quality of the generated solutions could be determined by a comparison to optimal solutions. However, as the problem is NP-hard, optimal solutions usually cannot be computed in reasonable time. Nevertheless, for a set of 1000 "small" (4 to 10 variables, access sequence length 50) instances of (1,0,1)-OA, for which optimal solutions have been computed, it has been shown that the GA achieved optimal results in all but two cases [Davi98]. To put this into context: For 10 variables, Liao's algorithm [LDK+95b] found the optimum only in 50 % of the cases, while the improved algorithm from [LeMa96b] achieved 80 %.

The GA has also been compared to two previous heuristic methods for offset assignment. Fig. 2.11 shows the results of a statistical comparison of the GA to the offset assignment technique with postpass MR optimization from [LeMa96b]. On the average, the GA generates solutions of 32 % lower costs.

Fig. 2.12 provides a comparison to Wess' technique [WeGo97a], which is based on *simulated annealing*. The results refer to an AGU configuration with 4 MRs and an auto-increment range of $r = 0$, i.e., only auto-modify operations are available. Wess proposed to *statically* load the 4 MRs with the values $\{-2, -1, 1, 2\}$, so as to simulate a $(k, 0, 2)$ AGU. The corresponding average results are represented by the left column (normalized to 100 %) in fig. 2.12. The center column shows the GA results for the *same* AGU configuration. The results do not differ much, since a simulated annealing approach may be expected to deliver results of similar quality. The GA achieves an average cost reduction of only 5 %. However, the advantage of the GA becomes obvious in

---

[4]All CPU times mentioned in this book refer to a Sun Ultra-1 workstation with 128 MB main memory and 100 MHz clock frequency.

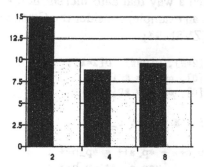

Figure 2.11. Results of GA-based $(k, k, 1)$-OA as compared to the heuristic from [LeMa96b]: The X-axis gives the value of $k = m$. The Y-axis gives the average offset assignment costs over a set of random problem instances (left column = heuristic, right column = GA).

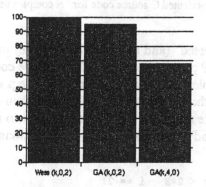

Figure 2.12. Results of GA-based offset assignment as compared to the simulated annealing approach from [WeGo97a]

the right column: Here, the restriction of static MR values has been removed, so that actually a $(k, 4, 0)$ AGU configuration with *dynamic* MR contents is exploited. This results in a cost reduction of 32 % on the average. Detailed experimental data can be found in appendix A (tables A.1, A.2, and A.3).

3. ADDRESS REGISTER ALLOCATION FOR ARRAY ACCESSES

3.1 MOTIVATION

A common characteristic of DSP algorithms is that most data processing takes place on data vectors inside loops. In C, data vectors are modeled as arrays, and the storage layout for arrays is fixed in the sense that neighboring array elements must be assigned to adjacent memory locations. One optimization that a compiler can perform for array accesses in loops is to allocate ARs to array accesses in such a way that auto-increment operations are exploited whenever possible. As an example, consider the C code in fig. 2.13, which is taken from DSPStone [ZVSM94].

```
int A[2*N], B[2*N], C[2*N], D[2*N];
int *p_a = &A[0], *p_b = &B[0];
int *p_c = &C[0], *p_d = &D[0];

for (i = 0 ; i < N ; i++, p_a++)
{
  *p_d   = *p_c++ + *p_a++ * *p_b++ ;
  *p_d++ -=          *p_a  * *p_b-- ;
  *p_d   = *p_c++ + *p_a-- * *p_b++ ;
  *p_d++ +=          *p_a++ * *p_b++ ;
}
```

Figure 2.13.   Pointer-oriented C source code for "N complex updates" benchmark

Such a "pointer-oriented" (and thus poorly readable) programming style is frequently used by DSP programmers in order to guide code generation in the compiler. In this example, four pointer variables (p_a to p_d) are used to access a pair of elements of the four arrays A, B, C, and D in each loop iteration. This programming style normally forces the compiler to allocate one AR for each pointer variable and to use auto-increment operations on ARs for address computations.

```
for (i = 0 ; i < 2*N ; i += 2)
{
  D[i]   = C[i] + A[i] * B[i] ;
  D[i]   += A[i+1] * B[i+1] ;
  D[i+1] = C[i+1] + A[i+1] * B[i] ;
  D[i+1] += A[i] * B[i+1] ;
}
```

Figure 2.14.   Rewritten C source code for "N complex updates" benchmark

A more abstract "array-oriented" C code for the same DSP routine is shown in fig. 2.14. In general, such a programming style should be preferred, since it results in better readable programs, and more freedom for optimizations is left to the compiler. Consider the accesses to array B in the loop body. The access sequence is

$$a_1 : \quad B[i]$$
$$a_2 : \quad B[i+1]$$
$$a_3 : \quad B[i]$$
$$a_4 : \quad B[i+1]$$

One obvious way to allocate ARs to the accesses would be to use one pointer p_1 for a_1 and a_3 and another pointer p_2 for a_2 and a_4. The pointers are initialized with &B[0] and &B[1], respectively, and both need to be incremented by 2 at the end of each iteration in order to point to the correct data pair in the next iteration.

If we want to optimize for performance, we may neglect the initialization cost and focus on the cost of address computations inside the loop body. Assume an AGU with an auto-increment range $r = 1$. When using the above scheme with two pointers, two extra instructions for address computations were required, since the increment of p_1 and p_2 cannot be implemented by auto-increment.

However, due to the special structure of the access sequence to array B, a better solution exists. All four accesses $a_1, \ldots, a_4$ may *share* the same AR, so that only a single pointer p is sufficient. Pointer p has to be initialized to point to &B[0] or access a_1 in the first iteration. Then, p has to be incremented for a_2, decremented for a_3, and incremented again for access a_4. Now p points to B[1]. In the next loop iteration, a_1 refers to B[2], so that at the end of the iteration p finally needs to be (auto-)incremented by 1 again. Using this addressing scheme requires only a *single* AR and *no* extra instructions for address computation inside the loop.

In fact it is exactly this scheme that has been used by the programmer of the code from fig. 2.13. For such small pieces of code it is comparatively easy to manually detect good solutions. However, in general this task is quite complicated and therefore should better be automated in the compiler. This can be illustrated by another small example. Consider the simplified C code in fig. 2.15, where we have seven accesses to array A in the loop body.

```
for (i = 2; i <= N; i++)
{
    A[i+1]    // a1
    A[i]      // a2
    A[i+2]    // a3
    A[i-1]    // a4
    A[i+1]    // a5
    A[i]      // a6
    A[i-2]    // a7
}
```

Figure 2.15.   Example array access sequence

If we assume that the loop step width L (1 in the example) is less or equal[5] to the auto-increment range r and if we allow for an arbitrary number of ARs, then

---

[5]We will restrict our considerations to this case in the following. If the assumption $L \leq r$ is not fulfilled, we can still use auto-modify operations for address computations, which however would complicate the

it is always possible to find an AR allocation for the array accesses, such that *all* address computations can be implemented by auto-increment. In this case, the primary optimization goal must be to *minimize* the number of required ARs. Minimizing this number maximizes the probability that the allocated "virtual" ARs can later be mapped to a limited number of physical ARs without the need for spill code.

There are two extreme solutions for AR allocation for the loop from fig. 2.15. If we use only a single AR, then the address computations for $a_1, \ldots, a_7$ are those shown in fig. 2.16. Only two out of seven computations are auto-decrement (a_1 and a_5), and five extra instructions are required in the loop body.

```
AR1 = &A[3];
for (i = 2; i <= N; i++)
{
    *AR1--       // a1
    *AR1 += 2    // a2
    *AR1 -= 3    // a3
    *AR1 += 2    // a4
    *AR1--       // a5
    *AR1 -= 2    // a6
    *AR1 += 4    // a7
}
```

Figure 2.16.   Addressing with a single AR

The second extreme solution allocates one AR per array access. In this case we get the code shown in fig. 2.17. All AGU operations are auto-increment, but seven ARs are required. This might well exceed the number of physically available ARs, especially if further arrays were referenced in the loop. In general, the optimum solution lies in between these two extremes, but it is not easy to find manually even for this simple example.

3.2   RELATED WORK

AR allocation techniques for array accesses have been considered by Liem [LPJ96]. He proposed a C-to-C transformation tool which replaces array accesses in the source code by accesses through pointers (cf. figs. 2.13 and 2.14). The sharing of ARs for multiple array accesses has been included, but has been based on some ad hoc rules. A more thorough investigation of this problem has been performed by Araujo et al. [ASM96], who presented a graph-based

---

discussion. We will also concentrate on array index expressions of the form "$i \pm c$" for some loop variable i and some constant c. More general index expressions, e.g. for multi-dimensional arrays, can mostly be transformed into this simple form by standard *induction variable elimination* techniques [ASU86].

```
       AR1 = &A[3];
       AR1 = &A[2];
       AR1 = &A[4];
       AR1 = &A[1];
       AR1 = &A[3];
       AR1 = &A[2];
       AR1 = &A[0];
       for (i = 2; i <= N; i++)
       {
         *AR1++  // a1
         *AR2++  // a2
         *AR3++  // a3
         *AR4++  // a4
         *AR5++  // a5
         *AR6++  // a6
         *AR7++  // a7
       }
```

Figure 2.17. Addressing with seven ARs

algorithm for optimized array index allocation. This technique will be presented in more detail in the next section. A different approach has been taken in [Gebo97a], where also physical AR limits have been incorporated into array index allocation. However, both techniques do not explicitly consider the address modifications required at the end of loop iterations.

3.3 OPTIMAL AR ALLOCATION IN LOOPS
PROBLEM FORMULATION

Consider a loop with a sequence of accesses $(a_1, \ldots, a_n)$ to some array A, where each a_i is of the form "$A[j + c_i]$", where j is the loop variable and c_i is an integer constant. We would like to solve the following optimization problem for an AGU with some auto-increment range r: Find an *AR allocation*

$$R : \{a_1, \ldots, a_n\} \to \mathbb{N}$$

such that

$$\forall 1 \leq i < j \leq n : \quad |c_i - c_j| > r \quad \Rightarrow \quad R(i) \neq R(j)$$

and

$$\max_{i=1 \ldots n} R(a_i) \to \min$$

In other words, function R allocates a minimum number of ARs for the array accesses, such that only auto-increment AGU operations are required inside the loop. Note that this formulation does not yet cover AR modifications at the end of an iteration, which are required for adjusting ARs for the *next* iteration.

Two accesses a_i and a_j may share an AR, only if their memory address distance does not exceed r. As proposed by Araujo [ASM96], opportunities for sharing ARs can be visualized by a directed *distance graph* $G = (V, E)$: Each node $a_i \in V$ represents the access a_i, and an edge (a_i, a_j) is in E, exactly if $i < j$ and $|c_i - c_j| \leq r$. Fig. 2.18 shows the distance graph for our example from fig. 2.15. For the purpose of illustration, we will assume $r = 1$ in the following.

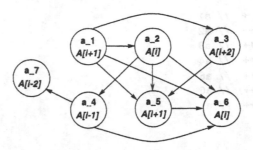

Figure 2.18.   Distance graph model for AR allocation

Consider the node representing access a_1. There are three possibilities for AR sharing with other nodes:

1. a_3, which requires an auto-increment

2. a_5, which requires no modification

3. a_6, which requires an auto-decrement

In general, sharing an AR between nodes means to follow a *path* in the distance graph G. For instance, (a_1, a_3, a_5, a_6) would be such a path in fig. 2.18. Since one AR has to be allocated for each access, any *complete* cover of G by a set of *disjoint* paths $(P_1, \ldots, P_m)$ represents a solution to the AR allocation problem. Furthermore, since the number of ARs must be minimized, a *minimum* number of disjoint paths is required.

MATCHING-BASED AR ALLOCATION

The relation between AR allocation and path covers in the distance graph has been observed by Araujo [ASM96]. He proposed to use an algorithm from graph theory [BoGi77] for solving the AR allocation problem by finding a *minimum disjoint path cover* of G. The algorithm works as follows:

Given a distance graph $G = (V, E)$, a *bipartite* graph $G' = (V', E')$ is built. For each $v \in V$, the node set V' contains two nodes v' and $\bar{v}$. For all $(u, v) \in E$, the edge set E' contains $(\bar{u}, v')$. On G' a *maximum cardinality matching* $Z \subseteq E$ is computed. Z is a maximum set of non-incident edges in E. Since G' is bipartite, computation of Z can be performed in $\mathcal{O}(|E'| \cdot \sqrt{|V'|})$.

It has been shown in [BoGi77] that an edge $e = (u, v)$ of the original graph G is part of a minimum disjoint path cover, if and only if the corresponding edge $e' = (\bar{u}, v')$ is contained in the matching Z. Thus, the optimal AR allocation can be directly derived from Z. Fig. 2.19 shows the result of applying this algorithm to our example. The optimum path set is

$$\{P_1 = (a_1, a_2, a_4, a_7), P_2 = (a_3, a_5, a_6)\}$$

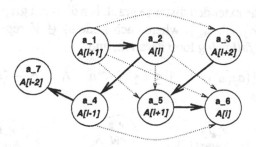

Figure 2.19. Matching-based AR allocation: Bold edges are selected for the minimum disjoint path cover.

The corresponding addressing scheme with two ARs is shown in fig. 2.20. Only auto-decrement operations are required for computing the memory ac-

```
AR1 = &A[3];
AR2 = &A[4];
for (i = 2; i <= N; i++)
{
    *AR1--      // a1
    *AR1--      // a2
    *AR2--      // a3
    *AR1--      // a4
    *AR2--      // a5
    *AR2 += 3   // a6
    *AR1 += 4   // a7
}
```

Figure 2.20. Addressing with two ARs

cesses *inside* an iteration, but the last two AR modifications which generate the addresses for the *next* iteration must be implemented by extra instructions. This is no surprise, since the problem formulation given above does not demand that also these address computations must have zero cost.

PATH-BASED AR ALLOCATION

In order to ensure that *only* auto-increment operations are required, the following additional constraint must be imposed on the mapping R: If P is a path in the cover of G representing an access subsequence $(a_{n_1}, \ldots, a_{n_k})$ and L is the loop step width, then $|c_{n_k} - (c_{n_1} + L)| \leq 1$ must hold. This means, it must be possible to generate the address for a_{n_1} in the *next* iteration from the address of a_{n_k} in the *current* iteration.

In [Leup97], we have proposed an *extended distance graph* to model this additional restriction. Let $G = (V, E)$ with $V = \{a_1, \ldots, a_n\}$ be the distance graph of a loop. The extended distance graph is a directed graph $G' = (V', E')$ with $V' = V \cup \{a'_1, \ldots, a'_n\}$, where each node $a'_i \notin V$ represents the array reference a_i in the *following* loop iteration, and

$$E' = E \; \cup \; \{(a_j, a'_i) \; | \; 1 \leq i \leq j \leq n \; \wedge \; |c_j - (c_i + L)| \leq 1\}$$

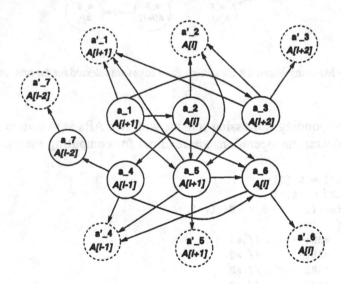

Figure 2.21.   Extended distance graph model

The extended distance graph for our example is shown in fig. 2.21. In order to satisfy the constraint $|c_{n_k} - (c_{n_1} + L)| \leq 1$, a minimum disjoint path cover in G' has to be found, where each node a_i must be touched by exactly one path. In addition, each path P that starts with node a_i must end in node a'_i because the same AR has to be used for a_i in all loop iterations.

Under this restriction, the above matching-based algorithm can no longer be applied, and in [Leup97] the following heuristic algorithm has been proposed:

1. Given a distance graph $G = (V, E)$, construct the extended distance graph $G' = (V', E')$, and assign a unit weight to each edge $e \in E'$.

2. Let a_i be the source node in $\{a_1, \ldots, a_n\} \subset V'$ with minimum index, i.e., there is no node a_j with $(a_j, a_i) \in E'$ and $j < i$. Compute the longest path $P = (a_i, a_{k_1}, \ldots, a_{k_m}, a'_i)$ in G' between a_i and a'_i.

3. Allocate a new AR for the array references represented by $\{a_i, a_{k_1}, \ldots, a_{k_m}\}$ in path P. Remove these nodes as well as the nodes $\{a'_i, a'_{k_1}, \ldots, a'_{k_m}\}$ from G', and remove all their incident edges.

4. If G' is not empty goto step 2, else return the number allocated ARs.

Computation of longest paths takes $\mathcal{O}(|V'| \cdot |E'|)$. In the worst case, each execution of step 3 removes only a single node pair (a_i, a'_i) from G'. Therefore, the runtime of the algorithm is bounded by $\mathcal{O}(|V'|^2 \cdot |E'|)$.

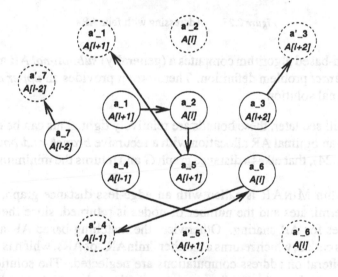

Figure 2.22.   Path cover in the extended distance graph

Fig. 2.22 shows the result (four disjoint paths) of this heuristic on the graph from fig. 2.21, and fig. 2.23 shows the corresponding addressing scheme. By construction, all address computations are auto-increment, but it is not guaranteed that the number of ARs (or paths in G') is minimal.

OPTIMAL AR ALLOCATION BY BRANCH-AND-BOUND

We now present an algorithm that actually finds the minimum number of ARs. The two algorithms described above have the following properties:

1. The matching-based algorithm computes an *optimal* AR allocation for a *relaxed* problem definition. Therefore, it provides a *lower bound* on the optimal solution.

```
AR1 = &A[3];
AR2 = &A[2];
AR3 = &A[4];
AR4 = &A[0];
for (i = 2; i <= N; i++)
{
    *AR1--    // a1
    *AR1++    // a2
    *AR3++    // a3
    *AR2++    // a4
    *AR1++    // a5
    *AR2      // a6
    *AR4++    // a7
}
```

Figure 2.23.   Addressing with four ARs

2. The path-based algorithm computes a (generally) *suboptimal* AR allocation for the *exact* problem definition. Therefore, it provides an *upper bound* on the optimal solution.

As we will see later, these bounds are relatively tight. This can be exploited to compute an optimal AR allocation with a recursive *branch-and-bound* algorithm (fig. 2.24), that takes a distance graph G and returns the minimum number of ARs.

If algorithm MINAR is called with an edge-less distance graph, then the recursion terminates and the number of nodes is returned, since there are no opportunities for AR sharing. Otherwise, the matching-based AR allocation algorithm is called, which returns a number "minAR" of ARs, which is minimal, if the inter-iteration address computations are neglected. The solution might *incidentally* require only auto-increment operations, in which case "minAR" can be immediately returned as the optimum. If the matching-based algorithm does not generate such a solution, the extended distance graph G' is built, and an upper bound "maxAR" is computed with the path-based heuristic.

Then, some edge e is selected from G, for which it is decided whether or not e must be part of the optimum path cover. Edge e must be *feasible*, i.e., if $e = (a_i, a_j)$ then a path from a_j to a'_i must exist in G'. Next, two graphs G_1 and G_2 are constructed. G_1 represents a partial solution in which e is *excluded* from the cover (by deleting e from G), while G_2 represents a partial solution in which e is *included* in the cover. This is implemented by merging nodes a_i and a_j in G.

Then, the matching-based algorithm is called twice to compute lower bounds "low1" and "low2" for G_1 and G_2. If G_1 requires at least a number of ARs larger than "maxAR" then excluding e from the path cover cannot result in an

```
algorithm MINAR
input: distance graph G = (V, E)
output: minimum number of ARs for G
begin
    if E = ∅ return |V|;
    end if
    minAR = MATCHINGBASEDALLOCATION(G);
    if matching-based allocation produced zero-cost solution then return minAR;
    end if
    G' = EXTENDEDDISTANCEGRAPH(G);
    maxAR = PATHBASEDALLOCATION(G');
    e = FEASIBLEEDGE(G, G');
    G₁ = EXCLUDEEDGE(G,e);
    G₂ = INCLUDEEDGE(G,e);
    low1 = MATCHINGBASEDALLOCATION(G₁);
    low2 = MATCHINGBASEDALLOCATION(G₂);
    if low1 > maxAR then return MINAR(G₂);
    end if
    if low2 > maxAR then return MINAR(G₁);
    end if
    return min(MINAR(G₁), MINAR(G₂));
end algorithm
```

Figure 2.24.   Branch-and-bound algorithm for AR allocation

optimal solution, and the optimum is $\text{MINAR}(G_2)$. Likewise, $\text{MINAR}(G_1)$ is the optimal solution, if the lower bound for G_2 exceeds the upper bound. In these cases, a part of the search space can be cut off without loss of optimality. In the worst case, the lower bounds cannot be used to prune the search space, so that the minimum has to be computed recursively for both G_1 and G_2.

Figs. 2.25 and 2.26 show the result of the branch-and-bound algorithm for our example array access sequence in the distance graph and the corresponding addressing scheme. As could be expected, the optimum lies between the lower bound (2 ARs, fig. 2.20) and the upper bound (4 ARs, fig. 2.23).

3.4   EXPERIMENTAL RESULTS

The worst case complexity of the branch-and-bound algorithm is $\mathcal{O}(2^{|E|})$. However, as lower and upper bounds can be computed efficiently, and as it is frequently possible to prune the search space early, the required runtime is reasonable in most cases. This is confirmed by a comprehensive statistical

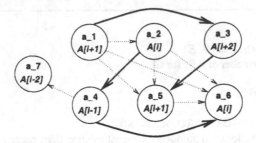

Figure 2.25.   Optimum path cover with three paths: (a_1, a_3, a_5), (a_2, a_4, a_6), and (a_7).

```
AR1 = &A[3];
AR2 = &A[2];
AR3 = &A[0];
for (i = 2; i <= N; i++)
{
    *AR1--      // a1
    *AR2--      // a2
    *AR1--      // a3
    *AR2++      // a4
    *AR1++      // a5
    *AR2++      // a6
    *AR3++      // a7
}
```

Figure 2.26.   Addressing with three ARs

evaluation of the three AR allocation algorithms, whose detailed results can be found in table A.4 of appendix A. Here, we only summarize the main results.

The average CPU time consumed by the branch-and-bound algorithm is in the range of a few seconds for access sequence lengths $n \leq 20$, while there is a jump to about 30 seconds for $n = 25$. Beyond this value, the path-based heuristic should probably be preferred in practice.

The average numbers of allocated ARs for different values of n are shown in fig. 2.27 for the matching-based algorithm (left columns), the path-based algorithm (center columns), and the branch-and-bound algorithm (right columns). The matching based algorithm performs surprisingly well with respect to the overhead of extra instructions for address computations, which are due to the neglect of inter-iteration AR modifications. On the average, less than one extra instruction is required.

This overhead is reduced to zero in the path-based heuristic, which requires more ARs. Naturally, the optimum number of ARs computed by the branch-and-bound algorithm is located between the lower and upper bounds.

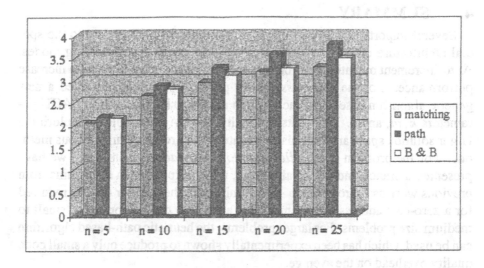

Figure 2.27.   Experimental results: AR allocation

As a by-product, the branch-and-bound algorithm allows to determine the quality of the path-based heuristic. For $n \leq 25$, an overhead less than 10 % has been measured. Thus the path-based algorithm provides a good means of efficiently computing close-to-optimum results for larger problem sizes.

We briefly mention an extension of this work, which has been described in [BLM98]. All three algorithms mentioned above minimize the number a of "virtual" ARs, so that only auto-increment operations are required in the loop body. However, if a exceeds the number k of physically available ARs, then the virtual ARs need to be "folded". Since each virtual AR corresponds to some path in the minimum disjoint path cover of the distance graph, folding two virtual ARs into a single physical AR means to merge two paths P_1 and P_2. Path merging in general implies the introduction of extra instructions for address computations. Let $P_1 = (a_1, a_3, a_5)$ and $P_2 = (a_2, a_4, a_6)$. The corresponding merged path is $P = (a_1, a_2, a_3, a_4, a_5, a_6)$. In the worst case, all address computations $a_1 \rightarrow a_2, a_2 \rightarrow a_3, \ldots$, now require an extra instruction, since a memory address distance less or equal r is no longer guaranteed. Therefore, in [BLM98] a *distance metric* for paths has been introduced, which measures the number of additional address computation instructions incurred when merging two paths. Using this metric, pairs of paths with a minimum distance are iteratively merged, until the physical register limit is met. Experiments indicated [BLM98] that this provides a good heuristic for mapping the minimum number of virtual ARs to physical ARs in case of $a > k$.

4.   SUMMARY

Several important code optimization problems for DSPs relate to the special architecture of AGUs, which provide auto-increment addressing modes. Auto-increment operations can be exploited to reduce code size and to increase performance. For the *offset assignment* problem, we have presented a new genetic algorithm based approach, which allows to handle arbitrary AGU parameters k, m, and r. Due to its generality, the GA is capable of exploring a larger solution space and thus also generates better results than previous methods. For the problem of *AR allocation for array accesses* in loops, we have presented a branch-and-bound algorithm, which exploits two algorithms from previous work as subroutines to *exactly* minimize the number of ARs required for a zero-cost addressing scheme. This algorithm can be applied to small to medium size problems. For larger problems, the heuristic path-based algorithm can be used, which has been experimentally shown to produce only a small code quality overhead on the average.

Chapter 3

REGISTER ALLOCATION FOR DSP DATA PATHS

This chapter deals with the *register allocation* problem in DSP code generation. Since DSP data paths typically show special-purpose registers, register allocation for such architectures has to be performed carefully in order to achieve high code quality. In particular, this holds if code is generated for *data flow graphs*, which are a widespread internal representation of program blocks. Such data flow graphs are composed of data flow trees, for which satisfactory code generation techniques have already been developed. In the following, we will extend this work by an algorithm that allocates registers for *common subexpressions* in data flow graphs, which can be considered as the interfaces between the data flow trees in the graph.

Fig. 3.1 shows, how register allocation fits into the overall compilation flow. In fact, the technique presented in the following also partially covers the code selection and scheduling phases, since for DSPs these phases can hardly be separated from register allocation.

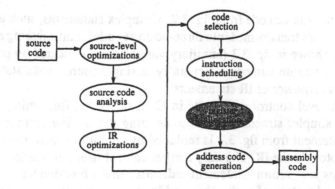

Figure 3.1.   Register allocation in the compilation flow

1.   BASIC DATA STRUCTURES

A common way to internally represent programs for code generation is to use *three-address code*. This names comes from the fact, that each three-address code statement refers to at most three operands. Since this representation is well-known from standard compiler construction [ASU86, Much97], we will not give a formal definition here, but merely exemplify how three-address code is used to represent more complex code, e.g. C source code. In the following, we will also use the term *intermediate representation* (IR) to denote a three-address code representation.

```
int f(int a, int b,int c)
{
  int x,y;

  x = a + b - 3 * c;

  if (x > 10)
  {
    y = x >> a + b - c;
  }
  else
  { y = x >> b - 10 * c;
  }

  return y;
}
```

Figure 3.2.   Example C source code

Consider the source code from fig. 3.2. Complex statements, such as x = a + b - 3 * c are transformed into three-address code by introducing *auxiliary variables*, as shown in fig. 3.3 (auxiliary variables are indicated by prefix "t", followed by a unique number). Generally, a single source code statement is replaced by a *sequence* of IR statements.

Also high-level control constructs in C (if-then-else, for, while, ...,) are replaced by simpler structures when generating the IR. For instance, the if-then-else statement from fig. 3.2 is replaced by conditional and unconditional jumps (or gotos) in the IR. The only further control structure needed in the IR besides jumps is a "return" statement indicating that a function has to be left.

The main advantage of such a three-address code IR is that code generation and optimization algorithms can work on a program representation with very

```
int f(int a,int b,int c)
{
  int x,y;
  int t1,t2,t3,t4,t5,t6,t7,t8,t9,t10;

  /* basic block B1 */

  t1 = a + b;
  t2 = 3 * c;
  t3 = t1 - t2;
  x = t3;
  t4 = x > 10;
  if (t4) goto L1;

  /* basic block B2 */

  t8 = 10 * c;
  t9 = b - t8;
  t10 = x >> t9;
  y = t10;
  goto L2;

  /* basic block B3 */

L1:
  t5 = a + b;
  t6 = t5 - c;
  t7 = x >> t6;
  y = t7;

  /* basic block B4 */

L2:
  return y;
}
```

Figure 3.3. Intermediate representation by three-address code. The syntax corresponds to the C-based IR generated by the LANCE system described in chapter 8.

simply entities instead of a source code containing complex statements, control structures and implicit address computations (for array and structure accesses).

The IR of each source code function can be decomposed into *basic blocks*. A basic block

$$B = (s_1, \ldots, s_n)$$

is a sequence of IR statements of maximum length that meets two conditions:

- The control flow enters B only at statement s_1, and

- the control flow leaves B only at statement s_n.

Thus, if statement s_1 is executed, then it is guaranteed the all other statements in B are executed as well. This is illustrated in fig. 3.3, where function "f" consists of four basic blocks. Given the IR of some function, it is easy to identify the basic block structure, since boundaries are explicitly given by jumps, return statements, and labels.

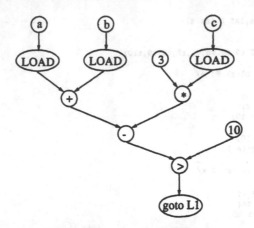

Figure 3.4. Data flow graph (DFG) representation of a basic block. Variables a, b, and c are supposed to reside in memory and are accessed via load operations.

For the purpose of code generation, it is convenient to visualize a basic block by a *data flow graph* (DFG). A DFG is a directed acyclic graph $G = (V, E)$. Each node $v \in V$ represents either a variable access, a constant, or an operation[1]. Each edge $e = (v, w) \in E$ denotes a *data dependence* between v and w, i.e., the result of the computation represented by node v is needed as an argument in operation w. In this context, we say that v *defines* a value *used* by w. Fig. 3.4 illustrates the DFG for the first basic block from fig. 3.3. As can be seen, the local variables in the IR become edges in the DFG model.

In fact, the DFG from fig. 3.4 represents only the special case of a *data flow tree* (DFT). We call a DFG a DFT, if it is connected and all nodes have a fanout less or equal to one. In general, a DFG needs not to be connected, and nodes may have a fanout larger than one. These nodes are called *common subexpressions* (CSEs), since they have multiple uses in the DFG.

2.   DATA PATHS WITH SPECIAL-PURPOSE REGISTERS

As already mentioned in chapter 1, most DSPs show *special-purpose registers*. From a processor architect's viewpoint, the use of such registers offers several advantages:

1. Special-purpose registers enable a kind of *pipelining* in the data path, which reduces the combinational delay and thereby allows for higher clock frequencies.

---

[1] Here, this term denotes all operations available in the C language.

2. For processors with instruction-level parallelism, the use of expensive multiple read/write ports for register files can be avoided.

3. The instruction word length is reduced, because registers are addressed implicitly.

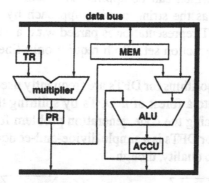

Figure 3.5. Data path of a TI 'C25 DSP: Special-purpose registers are TR (storing left multiplier inputs), PR (storing products), and ACCU (accumulator with feedback path to ALU)

An example DSP data path architecture with three special-purpose registers is shown in fig. 3.5. Other widespread DSP families (Analog Devices 210x, Motorola 56k) have even more different registers (or multiple small register *files*).

From a compiler designer's viewpoint, special-purpose registers are rather inconvenient, since the task of register allocation cannot be separated from code selection: Selecting a certain instruction for a DFG fragment implicitly binds values to registers, and vice versa. If code selection were performed without taking the detailed register architecture into account, then it might well be the case that a large overhead in code quality is introduced, because many register-to-register move instructions might need to be inserted in order to shuffle data to those registers where they are needed. Therefore, the concept of homogeneous virtual registers, combined with standard register allocation techniques (e.g. graph coloring) should not be applied for DSP code generation.

In addition, the small storage capacity of special-purpose registers has to be taken into account. In most cases, these registers (or register files) can store only one or two values at a time. Therefore, care has to be taken that registers are not overwritten before the last use of their respective values has taken place. Otherwise, expensive spill code would be necessary, and sometimes it is even impossible to directly spill a special-purpose register due to a missing transport path to memory. In the data path from fig. 3.5, for instance, spilling register TR is only possible via register PR (which in turn might contain a live value), and reloading PR from memory requires to perform a multiplication by 1.

3.   RELATED WORK

Optimal code generation for general DFGs, where the cost metric is induced by the cost values of the instantiated instruction patterns, has early been shown to be an NP-hard problem [BrSe76, AJU77]. However, this does not hold for DFTs, for which efficient algorithms do exist. Examples are the Aho-Johnson algorithm [AhJo76], which can be applied to machines with homogeneous register files, as well as the string parsing approach by Glanville [Glan77], where a linearized DFT representation is parsed w.r.t. a context-free grammar that represents the instruction set. Both require only linear time in the DFT size.

Code generation algorithms for DFTs are frequently used, because any DFG can be decomposed into a forest of n DFTs by splitting the DFG at its CSEs (fig. 3.6). Thus, replacing the code generation problem for a DFG by n code generation problems for DFTs is a simple divide-and-conquer approach, which may affect the solution quality, though.

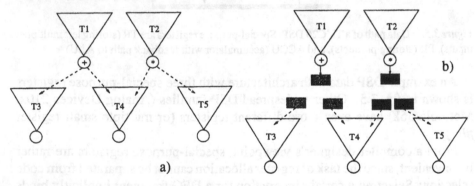

Figure 3.6. Splitting a DFG into DFTs: a) DFG with two CSEs (labeled with "+" and "*", triangles represent subtrees). b) Resulting forest of DFTs. Dedicated nodes (shaded and black boxes) have been inserted to represent the communication of CSEs between the DFTs.

3.1   TREE PARSING

The state-of-the-art technique in code selection for DFTs is *tree parsing*, usually implemented by *tree pattern matching and dynamic programming* [AGT89]. This technique offers four main advantages:

1. It is efficient, since it requires only linear time in the DFT size.

2. For a given instruction cost metric, it selects an optimum set of instruction pattern instances.

3. It is not restricted to homogeneous register machines and also captures *chained instructions* implementing multiple operations at a time (e.g. multiply-accumulate or complex addressing modes).

4. It is supported by publicly available tools, which is very important from a practical viewpoint.

In the tree parsing approach, the target instruction set is modeled as a *tree grammar*

$$G = (\Sigma_N, \Sigma_T, P, S, w)$$

where Σ_N is a set of *nonterminals*, Σ_T is a set of *terminals*, P is a set of *rules*, $S \in \Sigma_N$ is the *start symbol*, and $w : P \to \mathbb{R}$ is a cost metric for rules. We do not repeat the detailed theory of tree parsing here, since this has been extensively covered in the literature, e.g. [BDB90, FHP92b, FSW94, WiMa95].

Intuitively, the nonterminal set Σ_N is used to model hardware resources that can store data (registers, memories), while the terminal set Σ_T is used to represent operators and constants in a DFT. The grammar rules in P can be used to *derive* DFTs from the start symbol S. Like for usual string grammars, a derivation step in G means to replace the occurrence of a nonterminal $n \in \Sigma_N$ in a tree by another tree T, which is possible if the rule $n \to T$ is in P. The *tree language* $L(G)$ generated by grammar G is equal to the set of all DFTs, for which code might need to be generated.

The rules $p \in P$ are generally used to model the behavior of an instruction in the form of a small *tree pattern*. For instance, for an ADD instruction that computes the sum of two register contents and assigns the result to another register, the following rule would be used

$$reg \to PLUS(reg, reg)$$

where $reg \in \Sigma_N$ and $PLUS \in \Sigma_T$. Concerning the language *generated* by grammar G, this rule allows to *derive* a subtree with a root labeled *PLUS* and two subtrees from reg. Conversely, if we talk about tree *parsing*, the rule can be used to *reduce* a subtree rooted by *PLUS* and two subtrees (that have already been reduced to reg) to nonterminal reg. In any case, *using* the rule means to instantiate an ADD instruction.

For complex instruction patterns with common subexpressions on the right hand sides of rules (e.g. an expression specifying "base-plus-offset" addressing), nonterminals can also be used for *factoring* the grammar, so as to obtain a more concise model. In fact, there are usually large degrees of freedom in modeling an instruction set as a tree grammar, and finding the most suitable model requires some experience.

Since grammar rules essentially model instructions, the task of code selection for a DFT T is equivalent to finding a derivation in G for T from the start symbol S. Since the rules in P are weighted by function w, there are *optimal* derivations, i.e., derivations such that the sum over the weights $w(p)$ over all instances of rules p used in the derivation is minimal.

The tree parsing algorithm presented in [AGT89] constructs an optimal derivation for a DFT T as follows: In a *bottom-up* traversal, all nodes x in T are labeled with a set of triples (n, p, c), where $n \in \Sigma_N, p \in P$, and $c \in \mathbb{R}$. These triples represent the fact that node x can be reduced to nonterminal n at a total cost of c. The rule p implicitly determines the nonterminals, to which the subtrees of node x (if any) must be reduced in order to make p applicable for x. In general, multiple triples are annotated to x, which represent alternative derivations.

When the root of T has been reached, all alternative derivations potentially leading to an optimum are known. One optimum derivation is now explicitly constructed in a *top-down* traversal of T. For the root node, the triple (S, p, c) is selected (S is the start symbol), for which c is minimal over all alternative triples at the root. In turn, rule p now implies the optimum derivations for the subtrees at the next lower level in T, since the nonterminals to which they must be reduced are identical to the nonterminals on the right hand side of p. This traversal is recursively continued until the leaves of T have been reached and the derivation has been completely emitted.

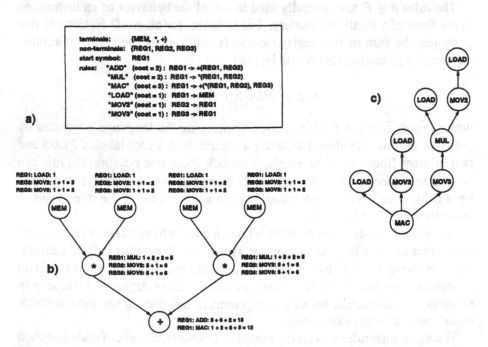

Figure 3.7. Example for tree parsing: a) Tree grammar specification, b) DFT with annotated nonterminal/rule/cost triples. There are two alternatives, ADD and MAC, for the root. MAC is selected, because it covers two operations at a time and therefore results in a cheaper derivation with a cost of 12 instead of 13 for ADD. c) Optimum derivation tree

This tree parsing method is illustrated in fig. 3.7. The algorithm requires *tree pattern matching* as a subroutine, in order to find rules matching a given node and possibly parts of its subtrees. The computation of the optimum derivation itself is an application of *dynamic programming*. The runtime is linear in the DFT size, since each node is visited only twice.

There are different tools supporting the automated implementation of tree parsing for a concrete instruction set, such as BEG [ESL89], TWIG [AGT89], IBURG [FHP92a], and BURG [FHP92b]. In OLIVE (an extension of IBURG) a tree grammar modeling an instruction set can be specified in Backus-Naur form. From the grammar specification, OLIVE generates a C source file containing a function for parsing an input DFT w.r.t. the grammar. The DFTs are represented by a C structure as shown in the following:

```
typedef struct tree {
    int op;
    struct tree* kids[2];
    int state_label;
} *DFT;
```

Here, component op is used to store the operation represented by a DFT node, array kids contains pointers to the children of a node, and state_label is a value internally used during DFT parsing. These are the minimum requirements for a DFT data structure when using OLIVE. The detailed structure may differ from the above example, since OLIVE accesses all DFT components via user-defined C macros. In particular it is possible to attach arbitrary application-specific information to DFT nodes by means of additional structure components, or to embed DFTs into a more general DFG data structure.

The generated C code can be compiled and linked to application programs. OLIVE offers two important improvements over IBURG which strongly facilitate the integration of OLIVE-generated code selectors into custom code generators.

1. The cost values $w(p)$ for grammar rules are not restricted to integer values, but arbitrary C functions returning values from an ordered set may be used as cost functions.

2. The grammar rules can be attributed with *action functions*. These are C functions performing some action (e.g. emission of assembly code) each time a rule is instantiated in the DFT derivation.

The code fragment from fig. 3.8 in OLIVE input syntax exemplifies these features.

Araujo [ArMa95] was the first to adapt tree parsing to DSP code generation. The key idea is the use of *register-specific* instruction patterns. In contrast to

```
%term AND                              // declare terminal AND

%declare<char*>reg;                    // declare nonterminal reg, whose
                                       // action function returns a string

reg: AND(reg,reg)                      // rule for a logical AND instruction
{
    $cost[0] = 1 + $cost[2] + $cost[3]; // cost = one plus cost of subtrees
}
=
{
    char* vr1, *vr2, *vr3;             // local variables in action function
    vr1 = $action[2];                  // get virtual register name for argument 1
    vr2 = $action[3];                  // get virtual register name for argument 2
    vr3 = NewVirtualName();            // get virtual register name for destination
    printf("\n AND %s,%s,%s",vr1,vr2,vr3); // emit assembly instruction
    return strdup(vr3);                // pass a copy of destination name upwards in tree
};
```

Figure 3.8.   Fragment of an OLIVE tree grammar specification

the above example, where only abstract registers are represented (nonterminal *reg*), one separate nonterminal is introduced for each special-purpose register, e.g. three in case of the TI DSP architecture from fig. 3.5. Using register-specific patterns allows to exactly model the argument and destination registers of instructions, as well as pure data move instructions. In this way, the tree parser already minimizes the amount of move instructions when selecting code for a DFT. Araujo also showed that for a certain class of DSP architectures, this approach actually generates optimal sequential assembly code for DFTs, even when instruction scheduling is taken into account, too. A similar technique has been developed independently by Wess [Wess00], however based on a completely different problem model.

3.2   CODE GENERATION FOR DFGS

In spite of its benefits, the major problem with the tree parsing approach is that still DFGs have to be split into DFTs, which strongly affects optimality. The reason is that, if code for a DFG is generated DFT by DFT, the CSEs have to be transported between the DFTs via a designated storage location L (represented by the boxes in fig. 3.6 b). This location L must be fixed, since if code is generated DFT by DFT, then each DFT must "know" in advance where its arguments (and also its result, if it happens to be a CSE) are located. Furthermore, L should not be a fixed special-purpose register, because then it would be very likely that L would be overwritten again before all uses of the CSE it stores have been scheduled. Therefore, the only "safe" CSE location L remaining for DSPs is the *memory*, which may store a large amount of live values at a time. It is obvious that communicating all CSEs via the memory in general implies an overhead in code quality.

Several attempts have been made to reduce this overhead. In [AML96], a DFG is partitioned into DFTs in such a way that architectural information about the target processor is already taken into account. For any CSE with u uses, this approach allows to read the CSE from a register in one use, while for the remaining $u - 1$ uses a reload from memory is required. The same holds for the code generation technique proposed in [LDK+95c].

Ertl recently generalized the tree parsing approach towards DFGs [Ertl99]. The required runtime is still linear, but optimality is only guaranteed for a certain class of tree grammars which primarily applies to processors with homogeneous register files. In contrast, fig. 3.9 visualizes the essential problem encountered in case of inhomogeneous (i.e. special-purpose) registers.

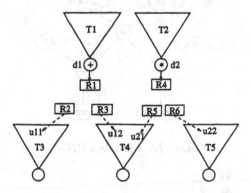

Figure 3.9.   Register allocation for CSEs with special-purpose registers

If we have a DFG with k CSEs, then there are k corresponding definitions $\{d_1, \ldots d_k\}$. Each d_i has a set of uses $\{u_{i1}, \ldots, u_{in_i}\}$. The definition and all uses have to be mapped to certain locations, and it might well be the case that all these locations are different in the optimal code.

In particular, it turns out that only using the memory as a CSE location L is not a good choice. In order to exemplify this, consider the subset of machine instructions of a TI DSP in fig. 3.10. These instructions can be executed on the TI 'C25 data path that has been shown in fig. 3.5.

Suppose, we must generate code for the DFG shown in fig. 3.11. Two alternative assembly programs are given in fig. 3.12. The code in the left column stores the CSE into a memory cell "temp". Since the CSE is a multiply operation, the result must reside in the product register PR. Thus, the contents of PR can be used again without the need to reload the CSE from memory. However, for the second use, the CSE must is reloaded from "temp". Such a solution would be generated e.g. by the algorithm from [AML96].

However, better (i.e., smaller and faster) code can be generated if both CSE uses are read from register PR. This is possible, since PR is not overwritten by the code generated for the first use. Thus, storing the CSE to memory can be

```
"lac":      ACCU = MEM
"addk":     ACCU = ACCU + CONSTANT
"add":      ACCU = ACCU + MEM
"pac":      ACCU = PR
"apac":     ACCU = ACCU + PR
"mpy":      PR = TR * MEM
"lt":       TR = MEM
"sacl":     MEM = ACCU
"spl":      MEM = PR
```

Figure 3.10.   TMS320C25 data path instructions

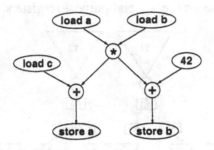

Figure 3.11.   Example DFG 1

```
lt a            lt a
mpy b           mpy b
spl temp        pac
pac             add c
add c           sacl a
sacl a          pac
lac temp        addk 42
addk 42         sacl b
sacl b
```

Figure 3.12.   Alternative assembly codes for DFG 1

completely avoided in this case. This saves one instruction (at least one, actually, since it might still be required to insert address computation instructions for spill code) and two memory accesses. Even though an instruction with a memory access has the same size as an instruction without one, instructions operating on registers should be preferred, since a memory access is generally slower than a register access. Also from a power consumption viewpoint, a register access will be typically cheaper.

The fact that register PR could be used in the above example is due to the special structure of the underlying DFG. If the first use of the CSE would take place in some DFT containing another multiply operation, then PR would be overwritten before the second use takes place. Thus, another register (or the memory) might be a better CSE location in this case. The decision where to optimally store a certain CSE cannot be made locally. This is easily seen when considering fig. 3.9 again: DFT T_4 has uses of two different CSEs, located in registers R_3 and R_5. Thus, the optimal code for T_4 depends on the locations R_1 and R_4, where the trees T_1 and T_2 place their results. In turn, the optimal choice of R_1 and R_4 depends on the code generated for T_3 and T_5, which have further CSE uses.

Eventually, it depends on the *entire* DFG, which location must be chosen for each CSE in order to produce good code. Therefore, a DFT-by-DFT code generation approach is less suitable for DSPs. In [BaLe99b] it has been proposed to perform code selection for DFGs while taking alternative CSE locations explicitly into account. However, this approach is comparatively difficult to implement, and moreover it still neglects another important problem: Not only does the optimal choice of CSE locations depend on the DFG structure, but also on the order in which the single DFTs are scheduled. This is shown in fig. 3.13.

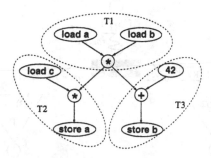

Figure 3.13.   Example DFG 2

Since DFT T_1 produces a CSE used in T_2 and T_3, T_1 has to be scheduled first. However, there is no such dependence between T_2 and T_3. If the CSE defined by T_1 is placed into register PR (this is useful, since multiply results are written to PR in any case), and T_2 is scheduled before T_3, then the contents of PR cannot be reused for T_3 because the multiply operation in T_2 overwrites PR. Conversely, if T_3 is scheduled before T_2, then PR can hold the CSE for both uses without getting overwritten. Thus, *alternative schedules* for the DFTs have also to be considered when allocating registers for CSEs. A technique which achieves this is presented in the following section. An alternative approach to the same problem has been proposed by Wess [Wess00], independently of the one presented here.

4.   REGISTER ALLOCATION FOR COMMON SUBEXPRESSIONS

4.1   PROBLEM DEFINITION

Given a DFG $G = (V, E)$ we would like to generate optimized *sequential* assembly code for G. The cost metric we use is the cost metric induced by the cost values $w(p)$ of the instruction patterns p, which might represent the size or the execution speed of an instruction.

The cost of a certain assembly program for G is the sum of the cost values of all required instruction pattern instances. Generation of sequential code, although possibly affecting global optimality, is a reasonable intermediate step in code generation, since instruction-level parallelism can usually be well exploited in a subsequent *code compaction* phase [DLSM81], which performs a partial re-scheduling of instructions.

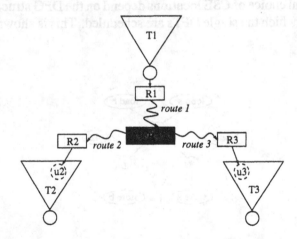

Figure 3.14.   Interface of three DFTs linked by a CSE

Since code generation for entire DFGs is a very complex problem, we will use a hybrid approach, which employs the efficient (and locally optimal) tree parsing technique for DFTs as a subroutine, while concentrating on register allocation for CSEs. This idea is illustrated in fig. 3.14, which "zooms" into the interface of DFTs linked by a CSE. DFT T_1 defines a CSE used by T_2 and T_3. The exact locations of the two uses u_2 and u_3 are DFT nodes representing operators.

When generating code for T_2 (e.g. by tree parsing) without considering the rest of the DFG, then there is a *locally optimal* location (a special-purpose register or the memory) R_2 for storing the argument of u_2, because u_2 is mapped to a certain functional unit which expects its operands in certain locations. Likewise, there is a locally optimal location R_3 for u_3. Also when considering

T_1, there is a locally optimal *destination* location R_1. In terms of tree parsing, the reduction of T_1 to nonterminal R_1 would have minimum costs[2].

Let $C(T)$ denote the cost value resulting when parsing a DFT T, and let $C(R_i \mapsto R_j)$ denote the costs of the sequence of instructions required to route a value from location R_i to R_j (zero if $i = j$). If we enforce that the CSE has to pass a certain location R, as illustrated in fig. 3.14, then the costs for the complete DFG are given as

$$C(G) = C(T_1) + C(T_2) + C(T_3) + C(R_1 \mapsto R) + C(R \mapsto R_2) + C(R \mapsto R_3)$$

Obviously, there is some location R which minimizes $C(G)$. However, as also the code generated for the DFTs themselves may depend on R, the selection of R should not be done locally for each CSE. Therefore, our problem formulation of register allocation for CSEs must be stated as follows: For a DFG G with k CSEs $\{c_1, \ldots, c_k\}$ and n DFTs $\{T_1, \ldots, T_n\}$, and a processor with m special-purpose registers $\{R_1, \ldots, R_m\}$, find a mapping

$$A : \{1, \ldots k\} \to \{1, \ldots, m+1\}$$

such that

$$\sum_{i=1}^{n} C(T_i^A) \to \min$$

$A(i) = m+1$ denotes that a CSE is assigned to memory instead of a register. The notation T_i^A means that DFT T_i has to be adapted to A as follows: If T_i generates some CSE c_j, then T_i must be reduced to the nonterminal representing location $A(j)$. This can be enforced by labeling the root of T_i with a special terminal. Conversely, if T_i uses c_j, then the corresponding leaf in T_i must be a terminal representing location $A(j)$. In this way we ensure that when parsing T_i the required data routing instructions are inserted automatically, and that the code generated for T_i actually depends on the location determined by $A(j)$.

4.2 REGISTER ALLOCATION BY SIMULATED ANNEALING

The total number[3] of possible solutions to the above register allocation problem is $(m+1)^k$. Since for each of these all $C(T_i^A)$ values have to be computed,

---

[2]Since the tree grammar must have a unique start symbol S, we assume that the grammar contains zero-cost rules $S \to R$ for any special-purpose register R (and also for the memory as a special location), so that DFTs virtually can be reduced to any register nonterminal.

[3]This number does even not capture the alternatives for scheduling DFTs (cf. fig. 3.13), which also influences the optimum mapping A. In our optimization approach the DFT schedule required to minimize the total costs is determined as a "by-product".

which takes at least linear time in the size of G, an exhaustive search is infeasible even for small problem sizes. We therefore use a *simulated annealing* (SA) algorithm [KGV83] to optimize the mapping A.

Similar to genetic algorithms (GAs), SA is suitable for nonlinear optimization problems, since it is capable of escaping from local optima in the objective function. The basic idea is to simulate a *cooling process*. Starting with an initial *temperature* and an initial solution, in each step the current solution is randomly modified. If the new solution is better, then it is accepted as the new current solution. Otherwise, it depends on the cost difference to the previous solution and the current temperature whether the new solution is accepted. During the annealing process, the temperature is lowered step by step, and the probability of accepting worse solutions decreases.

In contrast to GAs, which simultaneously store multiple alternative solutions, SA stores only a single solution at a time. Since for register allocation we are dealing with relatively large data structures (namely DFTs), we prefer SA in this case for sake of a more efficient implementation.

Fig. 3.15 shows the main algorithm. Similar to GAs, any concrete instance of an SA algorithm depends on a set of parameters that guide the optimization process. In our case, these are the initial temperature, the cooling factor, and the number of iterations in the inner loop. The concrete values shown in fig. 3.15 have been determined experimentally. We now describe the different subroutines.

DECOMPOSE: Performs the decomposition of the input DFG G into n DFTs by cutting the DFG at its CSEs.

INITIALCOST: Cost of the initial solution where all CSEs are assigned to the memory.

DOMODIFICATION: With equal probability, one out of two possible modify operations is chosen:

1. **Relocation of a CSE:** For two random numbers $r_1 \in \{1, \ldots, k\}$ and $r_2 \in \{1, \ldots, m+1\}$, the current CSE-to-register mapping A is changed by setting $A[r_1] := r_2$.

2. **Adding or removing a sequencing edge:** In order to take the effect of alternative DFT schedules into account, we use additional *sequencing edges* between DFTs in the graph G': We determine the order of code generation for DFTs by topological sorting (see below), while obeying *dependency edges* between DFTs defining and using CSEs (fig. 3.16). The dependency edges originally present in the DFG must not be changed in order to retain correctness of the code. However, these dependencies generally leave freedom for scheduling since they imply only a partial ordering of DFTs. Hence, we can explore different

```
algorithm CSE_REGISTERALLOCATION
input: DFG G with k CSEs;
output: sequential assembly code for G;
begin
   G' = DECOMPOSE(G);
   A[1..k] = m + 1; // assign all CSEs to memory
   current_cost = INITIALCOST(G');
   temp = 50;
   while temp > 0.1 do
      for count = 1 to 10 do
         DOMODIFICATION(G', A);
         schedule = TOPOLOGICALSORT(G');
         new_cost = 0;
         for all trees T in schedule do
            new_cost += COVERCOST(T);
         end for
         new_cost += ADDRCOST(schedule);
         if REGISTERCONFLICT(schedule) then new_cost = ∞;
         delta = new_cost - current_cost;
         if delta < 0 or RANDOM(0,1) < exp(-delta/temp)
            then current_cost = new_cost;
            else UNDOMODIFICATION(G', A);
         end if
      end for
      temp = 0.9 * temp;
   end while
   for all trees T in schedule do
      EMITASSEMBLYCODE(T);
   end for
end algorithm
```

Figure 3.15.   SA algorithm for CSE register allocation

alternative orderings by adding artificial dependencies represented by sequencing edges. Such a sequencing edge may be inserted only if the DFG remains acyclic. We can also remove a sequencing edge that has been inserted earlier in order to revise a scheduling decision. Whether a sequencing edge (and which) is added or removed is again determined by a random choice.

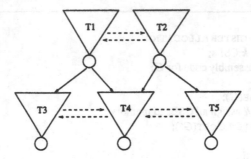

Figure 3.16.   DFG with two CSEs, dependency edges (solid arrows) and possible sequencing edges (dashed arrows)

TOPOLOGICALSORT: The partial ordering of the DFTs in G' is embedded into a total ordering compliant with the original dependency edges as well as with the current set of sequencing edges.

COVERCOST: For all DFTs in G', the cost of the corresponding machine code is computed by means of tree parsing. In our implementation, we use the OLIVE tool for this purpose. As opposed to a traditional code generation approach, where each DFT is parsed only once, tree parsing is slightly more complicated if we allow to dynamically change the location of CSEs, because actually this requires to dynamically change the underlying tree grammar. This is not directly supported by OLIVE, but we can use a modeling trick which exploits OLIVE's capability to specify arbitrary (in our case: *dynamic*) cost functions.

The tree grammar must allow to reduce any DFT to any register R, i.e., the result of the DFT is stored in R. Let S be the grammar start symbol. We use rules of the form

$$S \rightarrow \text{DEF\_CSE}(R)$$

for each special-purpose register R (and also for the memory). DEF_CSE is a dedicated terminal symbol that is attached to the root of each DFT defining a CSE c_i. When the tree parser determines a derivation of a DFT T_j from S, it must be ensured that T_j can only be reduced to $A[i]$. Thus, if $R = A[i]$, we set the cost value of the rule $S \rightarrow \text{DEF\_CSE}(R)$ to the cost value of the subtree already reduced to R. Otherwise, if $R \neq A[i]$, an infinite cost is assigned. Since we do not change the tree grammar itself, T_j could theoretically still be reduced to S. However, since the tree parser determines a minimum cost derivation, only that rule for which $R = A[i]$ can actually be selected.

Likewise, the uses of CSEs can be handled by dynamic cost functions in OLIVE. For each special-purpose register R, we define a grammar rule of the form

$$R \rightarrow \text{USE\_CSE}$$

USE_CSE is another special terminal denoting the use of a CSE c_i. The index i of that CSE can simply be attached as an attribute to the corresponding DFT node. If $R = A[i]$, then the rule gets zero cost, because in this case c_i is already present in R and just needs to be read out. Otherwise, the rule is assigned infinite cost, since c_i is only available in a location different from R.

ADDRCOST: In order to obtain an accurate cost value for the DFG G', not only the costs determined by tree parsing are taken into account, but also the additional costs possibly required for memory addressing instructions. For this purpose, we apply an offset assignment technique as described in chapter 2. Note that these addressing costs depend the current mapping A: The more CSEs are assigned to memory, the higher will be the addressing costs in general. Thus, placing a CSE in a register instead of the memory can eventually turn out to be more efficient, even though the costs determined by tree parsing may be higher. In addition, since the addressing costs depend on the memory access sequence, not only the amount of CSEs mapped to memory but also the detailed selection of those CSEs has an impact on the total costs. Therefore, in order to obtain a globally good solution, it is definitely favorable to include the computation of addressing costs in the inner loop of the optimization algorithm.

REGISTERCONFLICT: The above cost computation ensures that for any DFT a derivation is found which obeys the current CSE mapping A. Still, the solution might be invalid since it may require to overwrite a register currently storing a CSE whose uses have not yet all been scheduled. In order to detect this situation, sequential assembly code is generated for by means of Araujo's scheduling algorithm for DFTs [ArMa95]. For each CSE c_i, we decrement the number u_i of unscheduled uses each time a use has been scheduled. If some generated instruction writes a result to register R, where $A[i] = R$ and $u_i > 0$, then a register conflict is exposed, and the current solution is assigned an infinite cost value.

UNDOMODIFICATION: If the new solution is not accepted, then the previous solution is restored.

EMITASSEMBLYCODE: For the CSE mapping A and the DFT schedule resulting after termination of the outer loop, assembly code is emitted by a last run of the tree parser and the address code generator.

5.   EXPERIMENTAL RESULTS

The optimization algorithm described above has been experimentally evaluated for a number of basic blocks with CSEs, for which TI 'C25 (cf. fig. 3.5) assembly code has been generated. The basic blocks have been extracted from different benchmarks and DSP applications: DSPStone [ZVSM94], as well as C packages for GSM, JPEG [JPEG00], MPEG-2 [MPEG00], and MPEG-4 standards. The time for running the simulated annealing algorithm on a workstation typically ranges between 5 and 30 CPU seconds. Fig. 3.17 gives an overview of the main results. A table with detailed results can be found in appendix A in table A.5.

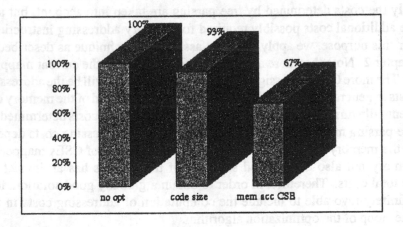

Figure 3.17.   Experimental results: CSE register allocation

The left column (set to 100 %) corresponds to solutions, where all CSEs are strictly assigned to memory. The center column shows the DFG costs (in terms of code size) when CSEs are allowed to reside in registers, as determined by the simulated annealing algorithm. We observed that on the average 38 % of the CSEs can be placed in registers. In total, this optimization resulted in a code size reduction of 7 % on the average.

The right column shows the results from a different perspective. Here, the number of memory accesses for CSEs have been counted, as compared to the initial solution with all CSEs in memory. Register allocation for CSEs resulted in an average reduction of those memory accesses to 67 %. This means a significant code quality improvement, when optimizing for power consumption and performance in case of using external memories.

The presented CSE register allocation technique can also be adapted to other DSP data paths. In a diploma thesis[4] [Bars99], it has been extended towards a TI 'C50 processor, a DSP similar to the TI 'C25 but with an additional accumulator buffer register. This buffer can be exploited to effectively store two CSEs at a time in the accumulator, which in turn gives higher degrees of freedom for CSE register allocation. An experimental comparison to Araujo's DFG code generation algorithm [AML96], which is not capable of exploiting this buffer, showed an average improvement in code size of 5.6 %.

6. SUMMARY

As DSPs typically show irregular data paths with special-purpose registers, dedicated register allocation techniques are required in order to avoid poor code quality. Previous work has shown how the standard tree parsing approach to code generation can be adapted for efficiently mapping DFTs to DSP data paths. However, DFGs with CSEs still cause problems, because CSEs need to be communicated between DFTs in such a way that the limited register file capacities are taken into account. A simple way to circumvent this problem is to pass all CSEs via memory, however at the expense of code quality losses. In contrast, we have pointed out that in principle any special-purpose register can be used to store CSEs in certain situations. The presented simulated annealing based technique performs an optimized mapping of CSEs to special-purpose registers for an entire DFG, while also taking alternative schedules into account. As a result, code quality improvements both in terms of code size and memory accesses have been measured. Also the offset assignment technique from chapter 2 nicely fits into this approach, which allows to partially couple the phases of code generation for DFGs and address code generation. From a software engineering viewpoint, it is important that existing code generator generator technology (OLIVE) can be reused to implement our approach.

---

[4]The author of that thesis also proposed an improved procedure for splitting DFGs into DFTs: Instead of breaking a DFG only at its CSE edges, one can also break "natural" edges, as defined in [AML96]. For a natural edge it is guaranteed that the corresponding value has to pass the memory due to the data path architecture itself. Thus, replacing the edge by memory accesses cannot reduce the code quality. However, the DFG is generally decomposed into more "small" DFTs, which in turn gives more opportunities for ordering the DFTs, so as to avoid register conflicts.

Chapter 4

INSTRUCTION SCHEDULING FOR CLUSTERED VLIW PROCESSORS

This chapter presents a code optimization technique for a special class of processor data paths, which we call *clustered VLIW*. Here, a cluster denotes a piece of the data path with functional units (FUs) and a local register file. Such a data path architecture is frequently found in multimedia processors and ASIPs. We will focus on the problem of instruction scheduling. Fig. 4.1 shows how this phase relates to the overall compilation flow.

Instruction scheduling for clustered VLIW data paths is complicated by the fact, that scheduling not only needs to aim at a good distribution of the workload across multiple FUs, but that the data transfers between local register files have to be taken into account, too. This resembles the problem of routing data between special-purpose registers that we considered in chapter 3. The difference in VLIW data paths, however, is that now we are dealing with general-purpose registers, except that registers are local to a certain FU. In addition, as

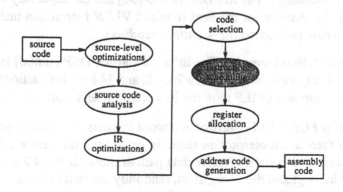

Figure 4.1. Instruction scheduling in the compilation flow

compared to a DSPs, a VLIW processor offers a large number of alternatives for binding instructions to FUs, since there are more FUs, and FUs are mostly multi-functional.

Our approach to constructing good schedules for clustered VLIW data paths is to *partition* instructions between clusters *during* the scheduling phase. In this way data transfers between clusters, which block FUs just like arithmetic operations, are inserted into the schedule dependent on the current resource availability. This is an example of phase coupling which, as we demonstrate experimentally, gives significantly better results than an approach using separate partitioning and scheduling phases.

1.   CLUSTERED VLIW PROCESSORS

A typical characteristic of a VLIW processor is that there are multiple FUs operating in parallel, which may execute instructions independently of each other. Due to this *instruction-level parallelism* (ILP), VLIW processors offer a very high peak performance, whenever the source program shows enough potential parallelism to keep the FUs busy. FUs not needed in a certain instruction cycle have to be set to an idle mode by "executing" a no-operation (NOP). This fact also implies a frequently mentioned disadvantage of VLIWs as compared to other processor classes: a very large code size. Nevertheless, the VLIW paradigm is increasingly used in recent DSP and multimedia processor designs, since processor architects have found ways to reduce this overhead:

Code compression: The Philips Trimedia [Phil00] has on-chip hardware for code compression and decompression.

Variable-length VLIW: The TI C6201 [TI00] uses variable-length instruction words to suppress NOPs. This is done by inserting "separator" bits into the instruction stream.

Differential encoding: The M3 DSP [FWD+98] has the capability of encoding only the *changes* for the next required VLIW instruction instead of a complete new instruction (*differential encoding*).

Multiple instruction formats: The Infineon Carmel DSP [Infi00] is capable of dynamically switching between 24, 48, and 144-bit instructions, dependent on the amount of ILP required in a program segment.

In case of n FUs, a VLIW instruction word consists of n (sub-)instructions We say that there are n *instruction slots*. In order to implement a set of fully orthogonal instruction slots, a VLIW data path as shown in fig. 4.2 a) would be required. It has a global file of registers, randomly accessible by each FU, and each FU is capable of executing all possible instructions. In this case the main hardware factor limiting performance is the number of FUs.

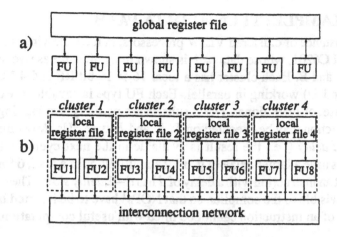

Figure 4.2.   VLIW data paths: a) orthogonal b) clustered

However, real-life VLIW processors differ from this model due to hardware complexity reasons: FUs are generally only capable of executing a certain subset of instructions, and register files may have only a small number of read/write ports. In fact, according to [FDF98b], multi-port register files are the most significant bottleneck in VLIW processors: Since a typical instruction performs 3 register accesses per cycle (reading 2 arguments, writing 1 result), an n-slot orthogonal VLIW needs a $3 \times n$-port register file. Many ports slow down performance, increase silicon area, and lead to wider instructions. In an extreme case, an n-slot VLIW may thus deliver a lower overall performance than a single-slot processor.

Therefore, a common compromise is to *cluster* the VLIW data path, as shown in fig. 4.2 b). There are multiple *local* register files (RFs), and each FU can communicate only with its own local RF. Each local RF, together with its FUs, is called a cluster. By clustering, the number of ports per RF can be largely reduced, at the expense of a *limited connectivity* between clusters. In fig. 4.2 b), this is indicated by an *interconnection network*. This (processor-specific) network enables the transfer of values between certain combinations of clusters. Any time a value produced in cluster i is needed in a different cluster j, a *copy operation* from RF i to RF j must take place.

A copy operation can be considered as a special type of instruction that uses an FU for performing an identity operation on its argument. Thus, any copy operation takes one instruction cycle and blocks one FU during that cycle. Therefore, it is obvious that an instruction scheduling algorithm for clustered VLIWs should take copy operations into account in order to generate high quality code.

2.   EXAMPLE: TI C6201 DATA PATH

As an instance of clustered VLIW processors, in the following we will consider the TI C6201, whose data path is shown in fig. 4.3. It has two symmetric clusters, A and B. Each cluster has a local 16×16 bit RF and 4 FUs (called L, S, M, and D) working in parallel. Each FU type is capable of executing a certain subset of operations. These subsets are partially overlapping, e.g., an ADD instruction may be executed on L, S, and D type FUs. Most instructions have a *unit delay* (i.e., the result is available in the next cycle), while some instructions have a larger delay (e.g. 3 for a multiply, 5 for a load, 6 for a jump). We say that an instruction with a delay of d has $d - 1$ *delay slots*. The instruction pipeline is visible to the compiler, so that NOPs[1] have to be inserted in case the delay slots of an instruction cannot be filled with useful computations.

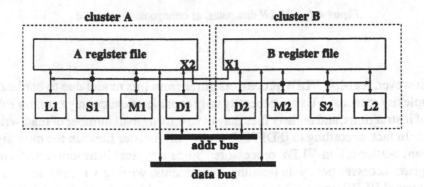

Figure 4.3.   Data path of a TI C6201 VLIW processor

The FUs in both clusters primarily work on their local RF. The only exception is that, in each instruction cycle, at most one of the units L, S, and M may read at most one argument from the opposite RF. Such read operations take place via the *cross paths* $X1$ and $X2$. Since there is only one cross path per cluster, at most two values can flow between A and B in each instruction cycle. Such a transport may be a usual copy operation from one RF to the other (via a MOVE instruction), but the value read over the cross path may also be directly consumed by some FU in the *same* cycle in which it is transported. In the latter case the value does *not* get stored in the local RF of that FU for further uses.

There are some other restrictions which need to be obeyed during instruction scheduling for the TI C6201:

---

[1] Here, a NOP denotes a full VLIW instruction, not a sub-instruction. For sake of smaller code size in case of multiple subsequent NOPs, the TI C6201 also has a multicycle NOP instruction which stalls the instruction pipeline for 1 to 5 cycles.

1. A copy operation is implemented by an addition of zero to the copied value. Although D units may execute additions, they cannot receive arguments from the opposite RF. Thus, only L and S units can be used for copy operations.

2. L units may receive either the left or the right argument from the opposite RF, while for S and M units the left argument *must* be read from the local RF.

3. Memory addresses (e.g. base plus offset) in LOAD/STORE instructions can only be computed on D units. An address computed on $D1$ in cluster A may be used to load (store) a value into (from) cluster B, and vice versa.

4. If two memory accesses are issued in parallel, then the two addresses (pointers) must be located in different RFs, and the two values values must be loaded into (or stored from) different RFs. For instance, two parallel loads into the RF of cluster A are invalid, even though there are two D units.

3.   PROBLEM DEFINITION

The scheduling problem we would like to solve can be stated as follows: Let B be a basic block, represented by an edge-weighted data flow graph $G = (V, E, w)$. According to fig. 4.1, we assume that code selection has already been performed, i.e., the DFG nodes in V represent concrete instructions, while DFG edges represent scheduling dependencies between instructions. Each edge e is weighted by an integer delay value $w(e)$.

We assume that the DFG nodes are not yet bound to one of the two clusters A and B. Thus, a *partitioning*

$$P : V \rightarrow \{A, B\}$$

of nodes between A and B must be computed. During scheduling it has to be decided, to which FU a node is bound, and at which point of time (or *control step*) its execution is started. For a given partitioning P, an *instruction schedule* is thus represented by two mappings

$$F : V \rightarrow \{L, S, M, D\}$$

$$C : V \rightarrow \mathbb{N}$$

We say that a schedule is *valid*, if for any node $v \in V$ the FU mapping is such that FU $F(v)$ belongs to cluster $P(v)$, $F(v)$ can implement the instruction represented by v, any FU is assigned at most one node per cycle, and the control step binding C does not violate any inter-instruction dependencies. The latter means for any node v with incoming edges

$$e_1 = (u_1, v), \ldots, e_k = (u_k, v)$$

the following constraint must hold:

$$C(v) \geq \max_{i=1}^{k}(C(u_i) + w(e_i))$$

The length $L(S)$ of a schedule $S = (F, C)$ is defined as the latest control step in which an instruction v, having a delay of $d(v)$, finishes its execution:

$$L(S) = \max_{v \in V} \ (C(v) + d(v))$$

Our goal is to *simultaneously* compute a partitioning P and a valid schedule (F, C) of minimum length. However, since resource-constrained scheduling is NP-hard even for a fixed partitioning [GaJo79], in practice we have to resort to a technique that generally produces only "close-to-optimal" solutions.

4. RELATED WORK

Scheduling algorithms can be divided into *global* and *local* algorithms. While local algorithms handle only a single basic block at a time, global algorithms work for an entire function.

Important global scheduling techniques include Trace Scheduling [Fish81] and Percolation Scheduling [AiNi88]. Their goal is to maximize performance of the *critical path* of a function across all its basic blocks. This is done by considering the critical path as a single large basic block (a *trace*), while temporarily neglecting the control flow. Since the block boundaries are removed, the critical trace can be scheduled more efficiently. However, scheduling an entire trace has an impact on the remaining basic blocks. Therefore, "repair code" has to be inserted, which may significantly increase the total code size.

An important global scheduling technique for program loops is *software pipelining*, which folds loop iterations so as to reduce the critical path length of the loop body [Lam88, JoA190]. This technique optimizes performance mainly for VLIW processors, but it can also be applied in ASIC synthesis [GVD89].

Global scheduling algorithms use local ones as subroutines. In our approach we will therefore focus on local scheduling. One important local scheduling algorithm, list scheduling, has already been described in chapter 1. We will use a variant of this technique in section 5.. Further standard local algorithms are *first-come-first-served*, which has been implemented e.g. in the MSSQ compiler [Nowa87], and the *critical path* method [DLSM81]. The latter possibly inspired the global Trace Scheduling approach, since it also gives priority to instructions on the critical path.

The problem with these standard techniques in the context of clustered VLIW processors is that they cannot perform instruction partitioning between clusters *during* scheduling. Either the partitioning (and the insertion of required copy operations) has to be done in advance, for instance based on a load balancing criterion [RKA99], or later in a post-pass phase. In any case, it is not sure that a close-to-optimum solution will result, because of the mutual dependence of scheduling and partitioning. In fact, one can easily observe that the optimum partitioning can only be determined *at the time of scheduling*: Instructions might be well balanced between the clusters, but the need to copy values may induce additional instruction cycles. In turn, free instruction slots in these additional cycles could be used to execute useful instructions. Thus, the insertion of copy operations generally has a global impact on the schedule.

Note that also min-cut graph partitioning techniques (e.g. the Kernighan/Lin heuristic [KeLi70]) are not useful here, because there is no need to *minimize* the communication between clusters. Instead, available communication resources and free instruction slots for copy operations should be fully exploited.

Code generation for clustered VLIWs has received research attention only recently. Nicolau et al. [CDN94] considered the problem of register allocation for VLIWs with multiple RFs. Their approach is based on a graph coloring technique, where the goal is to find a register allocation that meets the constraints imposed by the number of physically available RF read/write ports. Partial instruction rescheduling is performed in case that constraints are violated. However, the underlying VLIW processor model is quite abstract, so that special restrictions and the trade-off between copy operations and transfers via a cross path (as required in case of the TI C6201) are not considered.

In [JaVe99], a theoretical analysis has been provided which allows to derive lower bounds on the minimum schedule length for a fixed binding of DFG nodes to clusters. This technique could be used in a branch-and-bound algorithm for simultaneous partitioning and scheduling. However, due to some simplifying assumptions in the underlying processor model is it not clear whether this is also possible for real-life processors with special architectural restrictions, such as the TI C6201.

Fisher et el. [FDF98b] proposed a heuristic algorithm called *Partial Component Clustering* for the problem of partitioning DFG nodes between the clusters. The main idea is to assign subgraphs ("components") of the DFG to clusters in such a way, that copy operations along critical paths are avoided. The initial assignment is afterwards iteratively improved by swapping of component elements, while estimating the resulting schedule length with a simplified list scheduler.

A problem with this approach is that driving the partitioning phase by critical paths is mainly useful for DFGs, for which the critical path length is close to the actual minimum schedule length. In this case, nodes not lying on a critical path most likely can be scheduled in free instruction slots along the critical path (which is also the main motivation of the local critical path scheduling technique [DLSM81]). However, if there is a "wide" DFG, then the critical path length is only a very loose lower bound on the minimum schedule length, because the FUs become the limiting factor in scheduling. This can be shown by a small example for the TI C6201.

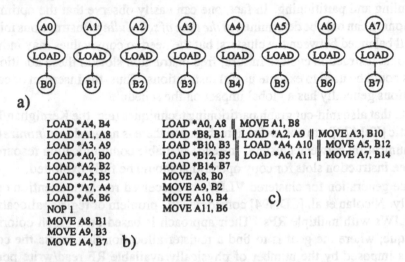

a)

LOAD *A4, B4	LOAD *A0, A8 ‖ MOVE A1, B8
LOAD *A1, A8	LOAD *B8, B1 ‖ LOAD *A2, A9 ‖ MOVE A3, B10
LOAD *A3, A9	LOAD *B10, B3 ‖ LOAD *A4, A10 ‖ MOVE A5, B12
LOAD *A0, B0	LOAD *B12, B5 ‖ LOAD *A6, A11 ‖ MOVE A7, B14
LOAD *A2, B2	LOAD *B14, B7
LOAD *A5, B5	MOVE A8, B0
LOAD *A7, A4	MOVE A9, B2
LOAD *A6, B6	MOVE A10, B4 c)
NOP 1	MOVE A11, B6
MOVE A8, B1	
MOVE A9, B3	
MOVE A4, B7 b)	

Figure 4.4.   Schedule length minimization by insertion of copy operations: a) data flow graph, b) schedule generated by TI C compiler (12 cycles, 12 instruction words), c) performance-optimal schedule (9 cycles, 16 instruction words), "‖" denotes parallel execution of instructions

In fig. 4.4 a), a simple-structured DFG is shown, containing 8 LOAD instructions. Each LOAD uses a register from cluster A as a pointer to load a value from memory into a register from cluster B. Since a LOAD has 4 delay slots and all LOADs are potentially parallel, the critical path length is 5. Fig. 4.4 b) shows the schedule generated by the TI C6201 assembly optimizer[2]. According to the restrictions mentioned in section 2., two LOADs can only be scheduled in parallel, if the pointers are located in different RFs. Since all pointers are initially located in RF A, the TI assembly optimizer generates fully sequential code[3]. The three MOVE instructions at the end could also be scheduled in parallel to earlier LOADs, but the schedule length of 12 would not be changed, since the 4 delay slots of the last LOAD instruction would then need to be filled with NOPs.

In contrast, fig. 4.4 shows a better schedule (in fact the schedule generated by the algorithm presented in the following section) with a length of only 9

---

[2]This tool is part of the software development toolkit for the TI C6201. It reads symbolic sequential assembly code (generated manually or by the TI C compiler) and performs partitioning, scheduling and register allocation.

[3]According to a personal communication, the TI assembly optimizer performs partitioning *before* scheduling, and therefore frequently does not find optimal or near-optimal solutions.

cycles. In the first 4 cycles, pointers located in A registers with an odd index are copied into RF B. In cycles 2 to 4, this allows to schedule two LOADs in parallel each. In cycles 6 to 9, the loaded values still residing in RF A are finally moved to their required locations in RF B. As can be seen, we have traded a larger code size for a faster schedule.

A simple analysis shows that the schedule from fig. 4.4 c) is performance-optimal: Obviously, it is impossible to issue the 8 LOADs within 4 instruction cycles or less, because at most two LOADs can be scheduled in parallel, and at least one MOVE of a pointer from A to B has to be executed before two parallel LOADs can take place. Thus, the LOADs must distributed over at least 5 instruction cycles. Suppose, the last LOAD is issued in cycle 5. Like any other LOAD, this LOAD instruction has 4 delay slots. Therefore, the DFG cannot be scheduled in less than 9 cycles.

5. INTEGRATED INSTRUCTION PARTITIONING AND SCHEDULING

In this section we present an algorithm that partitions instructions between clusters during scheduling, so as to achieve fast schedules. Since the algorithm is partially machine-specific, we will use the TI C6201 as a concrete target processor. The core of the technique, however, is certainly machine-independent and therefore could also be adapted to other clustered VLIW processors.

As already observed in [FDF98b], an exhaustive search for an optimal solution is clearly infeasible due to an extremely large search space. Therefore, as in chapter 3, we will rely on a simulated annealing (SA) algorithm. However, it would not be a reasonable approach to solve the entire partitioning and scheduling problem by SA, because for the scheduling problem itself good heuristics exist. In our case, we use a variant of list scheduling.

The main problem is, that performing list scheduling for a DFG requires that all instructions are already bound to clusters. On the other hand, we cannot determine a good partitioning before knowing the schedule length. Therefore, we perform partitioning and scheduling in an interleaved fashion: The partitioning is done by SA, while the scheduling algorithm serves as a cost function. In each iteration of the SA algorithm, our list scheduler uses a number of heuristics to obtain the best schedule possible for the current partitioning. The resulting schedule length is fed back to the SA algorithm to find a better partitioning.

5.1 PARTITIONING BY SIMULATED ANNEALING

The top-level procedure in our approach is the partitioning algorithm shown in fig. 4.5. It reads a DFG $G = (V, W, w)$ and returns a mapping $P : V \rightarrow$

$\{A, B\}$. We start with a random partitioning of the DFG nodes[4]. Variable "mincost" is used to maintain the current minimum schedule length. The initial costs are computed by the list scheduling algorithm described in the next section.

```
algorithm PARTITION
input: DFG G with N nodes;
output: P: array[1..N] of {A, B}; // partitioning
begin
    temp = 10;
    P := RANDOMPARTITIONING();
    mincost := LISTSCHEDULE(G,P);
    while temp > 0.01 do
        for i = 1 to 50 do
            r := RANDOM(1,n);
            P[r] := (P[r] = A) ? B : A;
            cost := LISTSCHEDULE(G,P);
            delta := cost - mincost;
            if delta < 0 or RANDOM(0,1) < exp(-delta/temp)
            then mincost := cost;
            else P[r] := (P[r] = A) ? B : A;
            end if
        end for
        temp = 0.9 * temp;
    end while
    return P;
end algorithm
```

Figure 4.5.   Partitioning algorithm

In each iteration of the annealing process, the cluster mapping of a random node is inverted. Then, the new costs are computed by another call to procedure LISTSCHEDULE for the new partitioning. As usual, better solutions are always accepted, while worse solutions are accepted at a rate inversely related to the temperature and the cost difference to the previous solution. If new new solution is not accepted, then the previous solution is restored by re-inverting the cluster mapping of the selected node.

---

[4]During experimentation we observed that using a heuristic seed for SA, where nodes are assigned to clusters in an alternating fashion, does not produce better solutions.

Finally, the solution found after termination of the outer loop is returned. Then, a last call of LISTSCHEDULE can be used to construct the corresponding schedule.

5.2   SCHEDULING FOR A FIXED PARTITIONING

The SA algorithm performs a kind of "blind" search, since it relies only on cost information delivered by the scheduler. Therefore, the scheduler must be "clever" and, under the restrictions imposed by the current partitioning P, try to achieve the best schedule possible. The scheduling main procedure (fig. 4.6) is a variant of the standard list scheduling algorithm. While not all nodes have been scheduled, it heuristically picks a node ready to be scheduled and inserts the node into the partial (initially empty) schedule constructed so far.

The heuristic used in NEXTREADYNODE is to select a node with a minimum ALAP (as late as possible) time. The rationale behind this is that the ALAP time is a metric for the "urgency" for a node to be scheduled next.

---

algorithm LISTSCHEDULE
input: DFG G, partitioning P;
output: schedule length;
var m: DFG node;
　　 S: schedule;
begin
　　 mark all nodes as unscheduled;
　　 $S := \emptyset$;
　　 while (not all nodes scheduled) **do**
　　　　 $m := \text{NEXTREADYNODE}(G)$;
　　　　 $S := \text{SCHEDULENODE}(S, m, P)$;
　　　　 mark node m as scheduled;
　　 end while
　　 return $\text{LENGTH}(S)$;
end algorithm

Figure 4.6.   Main scheduling algorithm

The "cleverness" of the scheduling algorithm lies in the procedure SCHEDULENODE (fig. 4.7), which uses a number of machine-specific heuristics to place a node into a partially constructed schedule. Such a schedule is represented as a list of control steps, each of which may contain a set of instructions already scheduled. The general strategy is to place the new node m into the earliest possible control step cs.

Function EARLIESTCONTROLSTEP first computes a lower bound for cs. If $n_1, \ldots, n_k$ are the predecessors of m in the DFG (possibly none), scheduled at control steps $C(n_1), \ldots, C(n_k)$, then the minimum cs value for m is given by

$$\max_{i=1}^{k}(C(n_i) + w(n_i, m))$$

Since the "repeat" loop executed next starts with an increment of cs, the initial cs value is set to EARLIESTCONTROLSTEP(m) - 1.

In the repeat loop, cs is iteratively incremented until a control step has been found into which m can be inserted without resource conflicts. First, an FU $F(m)$ in cluster $P(m)$ is determined (function GETFREEFU), which is capable of execution the instruction represented by m, and which is not yet blocked by another instruction at time cs. In case of alternatives, the choice is made arbitrarily, but the FU binding may still be revised later. If no free FU is directly found, then *version shuffling* [DLSM81] is applied to the current control step cs. Version shuffling tries to rearrange the FU binding of instructions already scheduled in cs, such that one FU capable of executing m gets free. If version shuffling fails to free a resource, then cs is incremented, and the insertion loop is repeated (indicated by the "continue" statement in fig. 4.7).

Without loss of generality we now assume that the instruction represented by m performs a *binary* operation with arguments x and y (scheduled in control steps $C(x)$ and $C(y)$, respectively), both of which reside in the cluster *opposite* to $P(m)$. The cases that m has less than two arguments or the arguments already reside in the RF of cluster $P(m)$ are simple special cases.

Since the clusters A and B are symmetric, we may also restrict our discussion to the case $P(m) = A$ and $P(x) = P(y) = B$. Thus, both values generated by x and y have to be transported from cluster B to cluster A. According to the TI C6201 interconnection network (cf. fig. 4.3), there are two possibilities[5] for this:

1. The transfer takes place in control step cs via cross path $X1$, which avoids a copy operation.

2. The transfer takes place via a copy operation scheduled earlier than cs.

We heuristically determine one of these alternatives, based on the current resource availability. The second case actually falls into two sub-cases. Therefore, function CHECKARGTRANSFERS checks three possibilities for both arguments (let z denote any of x and y).

---

[5] Theoretically, there is also a third possibility: copying a value via the memory. However, as both LOADs and STOREs have a significant delay, it is very unlikely that a benefit will result. Therefore, we neglect this option.

```
algorithm SCHEDULENODE
input: current schedule S, node m, partitioning P;
output: updated schedule S containing node m;
var cs: control step number;
begin
    cs := EARLIESTCONTROLSTEP(m) - 1;
    repeat
        cs := cs + 1;
        F(m) := GETFREEFU(m, cs, P);
        if F(m) = ∅ then continue; // try next cs
        if (m has an argument on a different data path) then
            CHECKARGTRANSFERS();
            if (at least one transfer impossible) then continue;
            else SCHEDULEARGTRANSFERS();
                if arg scheduling failed then continue;
            end if
        end if
    until (m has been scheduled);
    if (m is a LOAD instruction) then
        DETERMINELOADCLUSTER(m);
    end if
    if (m is a CSE with more than 2 uses) then
        INSERTFORWARDCOPY(S, m);
    end if
    return S;
end algorithm
```

Figure 4.7.   Scheduling algorithm for a single node

P1(z): If z is a *common subexpression* in the DFG, then it might be the case, that a copy operation for z from cluster B to A had already been scheduled in an earlier run of SCHEDULENODE for another use of z in cluster A. If such a copy operation does exist and happens to be scheduled in a control step less than cs then it **can be reused**.

P2(z): If there is a control step t in the interval $[C(z) + 1, cs - 1]$, where both $X1$ and one of the FUs $L1$ and $S1$ (the FUs capable of executing MOVE instructions) are free, then the transfer of z could take place via a **new copy operation** without increasing the schedule length.

P3(z): If $X1$ is free in cs, and the functional unit $F(m)$ is capable of reading z via $X1$ (dependent on whether z is a left or right argument of m), then a transfer of z **via the cross path** is possible.

If none of P1, P2, or P3 is possible at least for one argument, then the arguments cannot be provided to m in time, and the loop needs to be repeated with an incremented cs value.

Otherwise, function SCHEDULEARGTRANSFERS selects a combination of P1, P2, or P3 for x and y. While taking into account the architectural constraints of the TI C6201, the general strategy is to minimize the amount of resource blocking caused by the argument transfers. Another important concept here is to exploit *commutativity* of operations if possible. In particular, the following cases, ordered by priority, are checked:

P1(x) and P1(y): Copy operations are reused for both x and y. This is the best case, since it avoids blocking of any further resources.

P1(y) and not P3(x) and m is commutative: Normally, the transfer of x required a new copy operation, because no copy can be reused, and x cannot be transported via $X1$. However, if $X1$ is not yet blocked in control step cs, we avoid this by exploiting commutativity of m: The arguments are swapped, x is transported via $X1$, and a copy operation is reused for y.

P1(x) and P3(y): A copy operation is reused for x, and y is transported via the cross path. The first transfer does not block resources, while the second only blocks $X1$ in cs (a copy operation would block $X1$ *and* an FU). The case "P1(y) and P3(x)" is treated analogously.

P1(x) and P2(x): A copy operation is reused for x, and a new copy operation for y is inserted. The case "P1(y) and P2(x)" is treated analogously.

P3(x) and P3(y) and P2(y) and not P2(x): Both x and y could be transported via $X1$, a copy operation could be inserted for y, but no copy operation can be inserted for x. Then, $X1$ is preferred for x, while y is copied. The case "P3(x) and P3(y) and P2(x) and not P2(y)" is treated analogously.

P3(x) and P2(y): If m is not commutative, then x is transported via $X1$, and a copy operation is used for y. Otherwise, if m is commutative, the arguments are swapped, y is transported via $X1$, and a copy operation is used for x. This is a heuristic based on the following observation: If P3(x) holds, then the unit $F(m)$ must be an L unit, because L units are the only units capable of reading their left argument via a cross path. If we fixed x to $X1$, then the binding of m to L could never be revised again during version shuffling. This might be an unnecessary restriction when scheduling further nodes. The case "P3(y) and P2(x)" is treated analogously, except that we do not need to check for commutativity here, since m is not necessarily bound to an L unit.

P2(x) and P2(y): New copy operations are inserted for both x and y, if this causes no resource conflict. A resource conflict is exposed if the two copies could only be executed in the *same* control step, in which case a contention for (at least) $X1$ would occur. In this case function SCHEDULEARGTRANS-FERS fails to find a solution.

P3(x) and P3(y): Function SCHEDULEARGTRANSFERS also fails, because both arguments would need to be transported via $X1$ in the same control step.

If function SCHEDULEARGTRANSFERS fails, then the loop is repeated with an incremented cs value. Eventually, a valid insertion step cs for m is guaranteed to be found, because in the worst case we can always extend the schedule by new control steps. Node m is added to control step cs, and all required resources are marked as blocked.

Finally, two further heuristics are applied, which are generally beneficial for the scheduling of further nodes.

DETERMINELOADCLUSTER: If m represents a LOAD instruction, then the cluster to which the loaded value will be written, is not specified by the partitioning P. This is due to the fact, that memory addresses computed in one cluster can be used for loading a value into the RF of the opposite cluster. Only the cluster for computing the memory address itself is determined by P. This freedom can be heuristically exploited during scheduling. The architectural restriction that only one LOAD of a value into a certain cluster is valid per control step is modeled by two "virtual" resources Y_A and Y_B. For instance, if Y_A is already blocked in control step cs, then the destination cluster of the loaded value is set to B in order to enable two parallel LOADs, and vice versa. However, if both Y_A and Y_B are still free, then the value is loaded into that cluster, to which the majority of its uses are assigned by the partitioning P (fig. 4.8). The reason is that this decision will generally result in a lower number of copy operations. In case there is no such majority, the choice is made arbitrarily.

INSERTFORWARDCOPY: If m represents a common subexpression with more than two uses, m is not on a critical path, m is assigned to cluster A (B), and the majority of uses of m are assigned to B (A) then a copy operation $A \rightarrow B$ ($B \rightarrow A$) is inserted into the earliest possible control step after cs, even though no use of m has been scheduled so far. This heuristic generally enables a higher rate of reusing copy operations (case P1 mentioned above), which is illustrated in fig. 4.9.

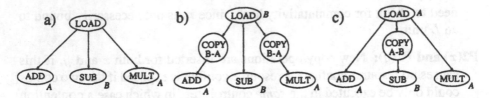

Figure 4.8. DETERMINELOADCLUSTER heuristic: a) Partial DFG: LOAD node with three uses mapped to clusters A, B, and A. b) If the value is loaded into cluster B, then two copies B → A might be necessary (unless one can be reused). c) If the value is loaded into cluster A, then at most a single copy A → B is required.

Consider the case that m is executed in step cs on cluster A, and m has two uses u_1 and u_2 on B. Suppose that u_1 is passed to routine SCHEDULENODE before u_2, and that a copy operation is required for the value transfer from m to u_1. Node u_1 will be scheduled in some step $C(u_1)$, while the copy operation will be scheduled in some step $t_1 \in [cs + 1, C(u_1) - 1]$. Now, consider the scheduling of node u_2: According to data dependencies and availability of FUs it might be possible to schedule u_2 before $C(u_1)$, but all transfer resources (cross paths and FUs for MOVE instructions) might be blocked. Placing u_2 into a control step $C(u_2) \geq C(u_1)$ allows to reuse the copy operation, but u_2 is scheduled unnecessarily late (fig. 4.9 a). However, if we had inserted a "forward" copy operation at an early control step, say $cs + 1$, then this copy could be reused both for u_1 and u_2, and u_2 could still be scheduled earlier than u_1 (fig. 4.9 b). Thus, the heuristic tends to achieve a lower number of control steps. However, we avoid inserting forward copy operations along a critical path. This is due to the fact, that any forward copy only *might* reduce the schedule length, while a copy operation along a critical path *guarantees* to increase the critical path length by one, which in turn increases the lower bound on the schedule length.

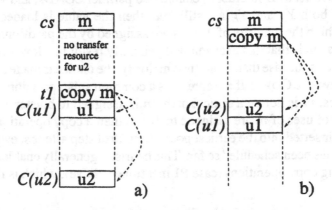

Figure 4.9. INSERTFORWARDCOPY heuristic

6.   EXPERIMENTAL RESULTS

In this section, we experimentally evaluate our technique by comparing the performance of generated schedules with schedules generated by the TI C6201 assembly optimizer.

6.1   STATISTICAL EVALUATION

For a broad evaluation of the above scheduling algorithm, first a statistical comparison to schedules generates by the TI assembly optimizer has been performed. Since the results are strongly influenced by the "width" of a DFG (cf. fig. 4.4), four sets of experiments have been carried out. We measure the width of a DFG by the ratio L/N, where L denotes the critical path length, while N denotes the number of DFG nodes. The DFG from fig. 4.4, for instance, has an L/N ratio of 5/8. An L/N value close to 1 indicates a "narrow" DFG, i.e., a DFG for which the minimum schedule length is close to L.

In particular for a VLIW processor, which allows to schedule many instructions in parallel, L is a tight bound on the minimum schedule length: Instructions on a critical path dominate the scheduling process, while instructions not lying on a critical path can just be scheduled in free instruction slots "along" that path.

In contrast, if $L/N \ll 1$, then the theoretical bound L is very loose, because in this case there are generally many nodes not lying on a critical path, which heavily compete for resources. In this case. the scheduling process is mainly dominated by limited availability of FUs and not by the critical path.

In our experiments, we therefore have considered four different sets of 100 randomly generated DFGs each, where the average L/N ratio of a DFG in its respective set has been 1.01, 0.62, 0.35, or 0.17.

Each DFG has been processed as follows: The internal DFG format has been converted into the sequential assembly code format read by the TI assembly optimizer. Then, this tool has been used to partition and schedule the sequential code. Additionally, the same DFG has been processed by the integrated partitioning and scheduling technique described above, and the resulting schedule lengths (in terms of instruction cycles) have been compared by a (custom) analysis tool. The maximum DFG size has been set to 100 nodes, for which the required CPU time is typically between 2 and 8 seconds.

The schedule length results for the four different L/N ratios are shown in fig. 4.10. As can be expected, there is a small difference for $L/N = 1.01$, because both schedulers were able to achieve the theoretical limit L in most cases. However, the difference becomes more significant with a decreasing L/N value. For "wide" DFGs ($L/N = 0.17$), our integrated approach on the average achieves a schedule length of 78 % as compared to the TI assembly optimizer. As already illustrated in the example from fig. 4.4, this improvement

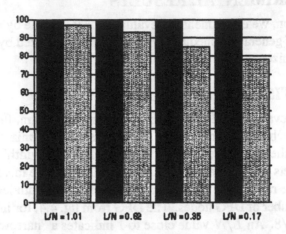

Figure 4.10.   Relative length of generated schedules: Left columns: TI assembly optimizer (set to 100 %), right columns: integrated technique

is due to the exploration of a larger search space and the better utilization of copy operations.

Fig. 4.11 shows the results from a different perspective, which highlights the influence of L/N on the difference between L and the actual minimum schedule length. For $L/N = 1.01$, both schedulers achieve the theoretical lower bound in most cases. In contrast, for $L/N = 0.17$, the TI assembly optimizer generates schedules of length $3.5 \cdot L$ on the average, while the integrated approach achieves approximately $2.5 \cdot L$.

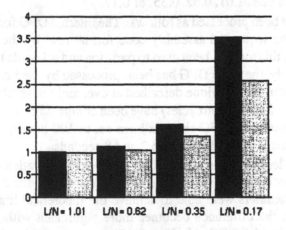

Figure 4.11.   Schedule lengths compared to lower bound L (corresponds to 1 on the Y-axis): Left columns: TI assembly optimizer right columns: integrated technique

The performance gain achieved by our integrated partitioning and scheduling technique has to be paid by an increase in code size. This is due to the fact, that the integrated technique generally makes more intensive use of copy operations, each of which requires one instruction word. We have measured the average overhead in code size as compared to the TI assembly optimizer for $L/N = 1.01$ and $L/N = 0.17$. In the first case, an overhead of 10 % has been observed, while in the latter case the overhead amounts to only 5 %. The reduced overhead for wide DFGs can be explained by the observation that also the TI assembly optimizer inserts a large number of copy operations in this case, which means that the code size significantly exceeds the theoretical lower bound N. Obviously, the integrated technique achieves a better utilization of the additional code words in the form of copy operations, while needing only slightly more such operations to achieve short schedules.

6.2   PERFORMANCE FOR REAL CODE

Fig. 4.12 shows performance results for a set of basic blocks extracted from realistic DSP programs. The left bars show the number of instruction cycles of machine code generated by the TI assembly optimizer, while the right bars show the corresponding results for our integrated scheduling technique. The benchmarks are ordered by increasing L/N ratio, ranging from 0.11 (left) to 0.61 (right). As predicted by the results of the above statistical evaluation, the performance gain tends to fall with increasing L/N ratio. The performance improvements compared to the TI scheduler range between 7 % (iir) and 26 % (dct). Thus, the statistical evaluation corresponds well with results obtained for realistic applications.

7.   SUMMARY

We have pointed out that for clustered VLIW processors, the tasks of partitioning instructions between clusters and instruction scheduling should be performed simultaneously in order to maximize performance. Our solution approach is to compute the partitioning by simulated annealing, while using a special list scheduler for determining a cost metric. This feedback path guarantees that the partitioning already takes into account detailed schedulability information.

In order to measure the code quality improvements for an existing machine, the list scheduler has been developed for the TI C6201 CPU with its special architectural restrictions. Porting the approach to another machine certainly requires to adapt the list scheduler. However, the approach in general is machine-independent. Our goal has been to demonstrate the optimization potential of phase-coupled partitioning and scheduling. In fact, the TI C6201 is a very interesting example for this purpose, because its capability of using cross paths for

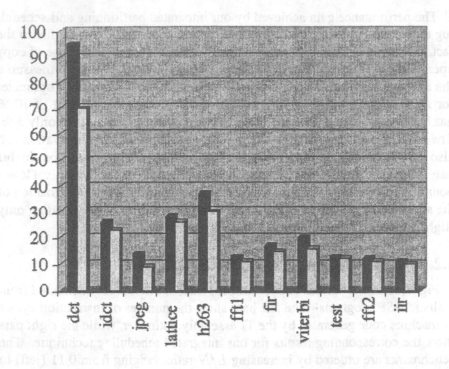

Figure 4.12.   Performance results for real DSP code

"volatile" copy operations adds another dimension to the search space. Since a good utilization of cross paths and regular copy operations requires a global view of the complete DFG to be scheduled, it seems unlikely that a human assembly programmer will be able to generate schedules of similar performance. This emphasizes the need for powerful code optimization techniques, that take into account the detailed architectural constraints of realistic VLIW processors.

Chapter 5

CODE SELECTION FOR MULTIMEDIA PROCESSORS

In this chapter we consider the problem of mapping a machine-independent intermediate representation into assembly code for multimedia processors. Fig. 5.1 shows the position of this code selection problem in the compilation flow.

We focus on the problem of exploiting SIMD (single instruction multiple data) instructions. These instructions, intended to increase the resource utilization, are comparatively difficult to handle by compilers. They perform multiple operations in parallel but, in contrast to usual VLIW instruction sets, SIMD instructions cannot be considered as being composed of separate "microinstructions". A simple mechanism offered by existing C compilers for multimedia processors is the use of *compiler-known functions*, which are similar to inline assembly code. Compiler-known functions allow to "call" SIMD instructions in C code, but still the responsibility of utilizing them is with the programmer. In addition, C code containing compiler-known functions is no

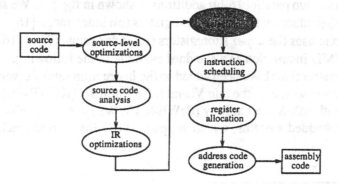

Figure 5.1. Code selection in the compilation flow

101

longer machine-independent. Therefore, code selection techniques are of interest, which are capable of exploiting SIMD instructions when compiling plain C source code. We present such a technique, and we demonstrate its applicability for two multimedia processors.

1.   SIMD INSTRUCTIONS

The SIMD paradigm, which is well-known in computer engineering, denotes the idea of speeding up program program execution by performing the *same* type of computations on *different* data in parallel. This concept, also called vectorization, has been recently adapted to the instruction level for multimedia processors. As already mentioned in chapter 1, these processors are intended as uniform, high-performance platforms for multimedia applications.

Different types of media data require different arithmetic precisions. For instance, audio data are frequently encoded as streams of 16-bit samples, while image data often require a word length of only 8 bits. In addition, also general-purpose routines, e.g. for control functions, must be executable on multimedia processors, for which word lengths of 32 or 64 bits are common.

Executing computations on 8 or 16-bit data on a data path with a much larger word length obviously implies a waste of computational resources. Therefore, multimedia processors typically allow to "split" the data path, while performing independent (but identical) computations on certain parts of the argument registers and also writing the results to certain parts of the destination register. Fig. 5.2 illustrates the use of registers in this case for a 32-bit machine. If we assume the bit widths 32, 16, and 8 are used for integer, short integer, and character data, respectively, then (in terms of the C language) any register may hold 1 "int", 2 "shorts", or 4 "char" data at a time. In the latter two cases, we say that a *full* register is split into (2 or 4) *subregisters*.

We call an instruction a *SIMD instruction*, if it performs a set of identical computations on subregisters in parallel[1]. An example for a SIMD instruction that performs two parallel 16-bit additions is shown in fig. 5.3. We say that one addition takes place on the *lower* subregisters (bit index range $[15 \ldots 0]$), while the other one uses the *upper* subregisters (bit index range $[31 \ldots 16]$). We will use this SIMD instruction as a standard example in the following.

SIMD instructions have been added to the instruction sets of several general-purpose processors, e.g., the Sun Visual Instruction Set [KMTP+95] and Intel's Pentium with MMX Technology [PeWi96, PWW97]. Multimedia processors used in embedded systems, which support SIMD instructions include the TI

---

[1] Other terms for this feature in the literature are *split-ALU instructions, short-vector instructions, subword parallelism,* or *SIMD within a register (SWAR)*. However, the term "SIMD instructions" appears to become common, see e.g. Intel's WWW pages (www.intel.com).

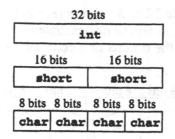

Figure 5.2.   Splitting of 32-bit registers into multiple subregisters

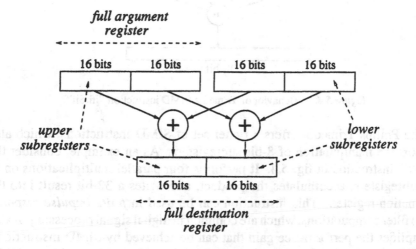

Figure 5.3.   Example SIMD instruction "ADD2"

C6201 [TI00] and the Philips Trimedia TM1000 [Phil00]. Both are 32-bit machines.

The TI C6201 offers two "native" SIMD instructions, called ADD2 (which corresponds to fig. 5.3) and SUB2 (like ADD2, but performing two subtractions). With respect to multiplication, the TI C6201 is restricted to 16×16 bit, generating a 32-bit result. Therefore, it has four types of multiply instructions. Let D be a destination register, and let $R_1 = S_{1,up} \circ S_{1,lo}$ and $R_2 = S_{2,up} \circ S_{2,lo}$ be two argument registers, each composed of two subregisters (denoted by the concatenation operator "$\circ$") $S_{i,up}$ and $S_{i,lo}$. Then, the four multiply instructions behave as follows:

MPY: $D = S_{1,lo} \times S_{2,lo}$
MPYLH: $D = S_{1,lo} \times S_{2,up}$
MPYHL: $D = S_{1,up} \times S_{2,lo}$
MPYH: $D = S_{1,up} \times S_{2,up}$

These instructions are not actually SIMD instructions, and the capability of performing a "cross-wise" multiplication of subregisters by MPYLH and MPYHL is usually used to emulate 32×32 bit integer multiplications by a sequence of instructions. However, as will be demonstrated later, MPYLH and MPYHL can also be exploited in the context of a code selection technique for SIMD instructions.

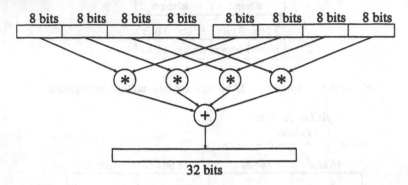

Figure 5.4.   Behavior of Trimedia SIMD instruction "ifir8ii"

The Philips Trimedia offers a richer set of SIMD instructions, which also support the manipulation of 8-bit subregisters. As an example, consider the "ifir8ii" instruction in fig. 5.4. It performs four parallel multiplications on 8-bit subregisters, accumulates the product, and writes a 32-bit result into the destination register. This instruction can be used in *finite impulse response* (FIR) filter computations, which are common in digital signal processing. It also exemplifies the performance gain that can be achieved by SIMD instructions: An FIR computation can be executed four times faster than without SIMD instructions.

Arithmetic SIMD instructions require special hardware support. For the above ADD2 instruction, for instance, carry propagation in the ALU has to be suppressed after the 16th bit. In contrast, some 32-bit instructions can be employed as SIMD instructions as well without special hardware. First, this concerns bitwise logical instructions (AND, OR, NOT, ...). For instance, performing a 32-bit AND on full registers can also be interpreted as performing two parallel ANDs on two pairs of 16-bit subregisters. Second, also 32-bit LOAD and STORE instructions can (and in fact should) be used in the context of SIMD instructions. This can be demonstrated with the following C code fragment:

```
void f(short A[],short B[],short C[])
{ int i;
  for (i = 0; i < N; i += 2)
  { A[i]   = B[i] + C[i];
    A[i+1] = B[i+1] + C[i+1];
  }
}
```

This code performs an addition of two vectors B and C containing 16-bit ("short") elements. The for-loop has been unrolled once, so as to exhibit the potential parallelism. Obviously, an ADD2 instruction (see fig. 5.3) can be used to implement the two additions in the loop body in parallel. This requires that $B[i]$ and $B[i+1]$ are located in the upper and lower subregisters of some full register R_1, while $C[i]$ and $C[i+1]$ must reside in the upper and lower subregisters of another full register R_2. Two 32-bit LOADs can be used to satisfy this precondition. Since $B[i]$ and $B[i+1]$ are stored in adjacent memory locations, a single 32-bit LOAD can be used to simultaneously fetch both data from memory into a register[2], and the same holds for $C[i]$ and $C[i+1]$. After execution of ADD2, a single 32-bit STORE suffices to write the two results $A[i]$ and $A[i+1]$ back into memory. This is illustrated in fig. 5.5.

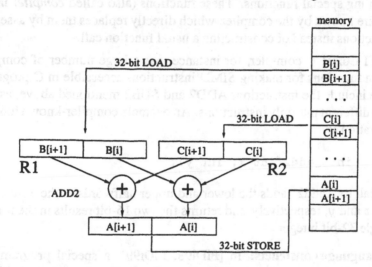

Figure 5.5. Parallelization of vector addition with SIMD instructions

2. RELATED WORK

The implementation of SIMD instructions to support execution of multimedia applications is a quite recent idea. It is mainly motivated by the large potential performance gain, which also has been experimentally studied [BJER98, LeSt98, RAJ99].

However, current C compilers for multimedia processors cannot adequately exploit SIMD instructions[3]. This means a potentially large code quality loss,

---

[2]On some processors, this requires a certain memory alignment, which we will neglect here for sake of simplicity. Any required alignment can usually be enforced by special compiler or assembler directives.
[3]In fact, most compilers do not make use of SIMD instruction at all. According to recent announcements, there are some first compiler releases for Intel's MMX capable of automatically using SIMD instructions for

if application are written in "plain" C code. Three methods of circumventing this problem are currently in use:

Assembly code libraries: Critical routines which qualify for the use of SIMD instructions are implemented by hand-written assembly code. For the Intel Pentium MMX, for instance, assembly code libraries exploiting SIMD instructions are publicly available [Inte00a].

Also direct inline assembly code can sometimes be used in C programs. However, it is well-known that inline assembly is to be used very carefully, since it may strongly affect the code generation process in the compiler.

Compiler-known functions: SIMD instructions can be instantiated in C code by calling special functions. These functions (also called *compiler intrinsics*) are known by the compiler, which directly replaces them by assembly instructions instead of constructing a usual function call.

The TI C6201 C compiler, for instance, has a large number of compiler-known functions for making SIMD instructions accessible in C programs. These include the instructions ADD2 and SUB2 mentioned above, as well as the different multiply instructions. An example compiler-known function declaration is

```
int _add2(int x, int y);
```

Any call to "_add2" adds the lower and upper halfwords of the 32-bit variables x and y, respectively, and returns the two 16-bit results in the form of a single 32-bit integer.

Special language constructs: In [FiDi98, FiDi99], a special programming language has been proposed, which allows for a source-level vectorization of operations. The corresponding compiler generates C code as output, which contains calls to compiler-known functions. This increases the programming comfort, however at the expense of introducing another programming language.

Intel's C++ compilers for the Pentium MMX have a *vector class library*, using which the vectorization of statements can also be made explicit in the source code [Inte00b].

Even though the above methods give access to SIMD instructions from the C level, there are two significant problems: In all cases, one has to use machine-specific code, and a good utilization of SIMD instructions still has to be ensured

---

simple loop bodies (e.g. like our above vector addition example). As we will show, the technique we present is more general, since it does not require a special data flow graph structure.

by the programmer. A compiler capable of automatically exploiting SIMD instructions for machine-independent (and thus portable) C code would obviously be more favorable. Our goal in the remainder of this chapter is to develop an approach to solving this problem.

Using SIMD instructions in code generation is mainly a matter of code selection, during which intermediate code is mapped into assembly instructions. Obviously, classical code selection techniques cannot be directly used with SIMD instructions (otherwise current compilers would definitely exploit them). The reasons are explained in the following.

In chapter 3 we have already summarized the main features and limitations of *tree parsing* as a code selection technique. The general idea is to decompose data flow graphs (DFGs) into data flow trees (DFTs), and to cover[4] DFTs separately by instruction patterns, each of which is represented by a "small" tree . As can be seen in fig. 5.3, however, SIMD instructions are not tree-shaped, but they generally implement multiple "small" trees at a time.

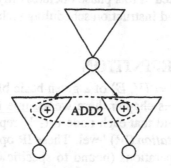

Figure 5.6. DFG decomposed into three DFTs. Exploiting an ADD2 instruction may require to cover two potentially parallel "+" nodes located in different DFTs.

Therefore, code selection with SIMD instructions requires a more global view of the entire DFG, which is illustrated in fig. 5.6. There are different options of generating SIMD instructions during code generation:

1. Operations on subregisters are temporarily considered as valid "stand-alone" instructions. Then, code selection can be performed as usual (i.e. DFT by DFT), and selected operations on subregisters are later combined to SIMD instructions during register allocation and instruction scheduling.

2. SIMD instructions are already generated when covering the DFG, while considering multiple DFTs at a time.

---

[4]In the following we occasionally use the more intuitive term "covering" to denote the derivation of a DFT from the start symbol of a tree grammar.

The first option is simple from a code selection viewpoint. However, the phases of register allocation and instruction scheduling get very difficult. The standard graph coloring based register allocation approach is based on lifetime analyses of virtual registers, and it is assumed that each virtual register occupies one physical register. This gets more complex if virtual registers are subregisters: Not only must the register allocator decide, which virtual registers with non-overlapping lifetimes will share a physical register, but also which virtual registers will *simultaneously* be present in a physical (full) register. The latter in principle means to couple the lifetimes of virtual registers, which makes register allocation a heavily constrained problem. In addition, during instruction scheduling one might find that a valid packing of operations on subregisters to SIMD instructions is not possible, for instance due to scheduling deadlocks. In turn, this might require time-consuming backtracking.

Overcoming these difficulties might be possible, but we prefer the second option, which appears to be simpler. It requires more effort during code selection, but since the code generated in this phase operates only on full registers, standard register allocation and instruction scheduling techniques may afterwards be applied.

3.   PROBLEM DEFINITION

We consider a DFG $G = (V, E)$ of a given basic block. In contrast to the DFGs used in the previous chapter, here we assume that code selection has not yet been performed and that each node $v \in V$ represents an operation at the *intermediate representation* (IR) level. These IR operations are arithmetic, logic, and comparison operations (bound to specific data types), as well as LOADs, STOREs, constants, and jumps[5]. The DFG edges $e \in E$ represent data dependencies between IR operations. Additional forms of dependencies, i.e., control dependencies, anti-dependencies, and output dependencies need not to be considered in this phase, but they are obeyed in the second phase.

We would like to determine an optimal *DFG cover* by a set of instruction pattern instances, where the cost metric is induced by the cost metric of the single instruction patterns. By *covering*, we denote the selection of a set of instructions operating on virtual registers, which (after register allocation and instruction scheduling) implement the computation represented by the DFG.

Similar to the technique presented in chapter 3, our approach uses *tree parsing* as a subroutine. However, the context is very different, since in case of multimedia processors we do not have to deal with special-purpose registers. Thus, cutting a DFG into DFTs at its common subexpressions (CSEs) and passing the CSEs via the general-purpose registers generally does not cause

---

[5] A more detailed definition of an intermediate representation suitable for the C language will be provided in chapter 8.

a significant overhead in code quality[6]. Instead, the main problem is that the tree parsing technique inherently cannot exploit any form of instruction-level parallelism. Therefore, it cannot be directly applied to SIMD instructions.

The code selection technique we propose consists of two separate phases: In phase one, a DFG is covered by instruction patterns, while keeping alternative solutions. This phase only considers one DFT at a time. Whether or not a certain node is covered by a SIMD instruction is decided only during the second phase which, based on the available alternatives, ensures an optimal exploitation of SIMD instructions across an entire DFG. These phases are described in the following two sections. For sake of an easier explanation, we focus on the case of a full 32-bit register subdivided into two 16-bit subregisters. However, the technique can be easily scaled to the case of four 8-bit subregisters. Section 6. describes the necessary extensions.

4.   GENERATION OF ALTERNATIVE COVERS

A given DFG G is decomposed into n DFTs $T_1, \ldots, T_n$ by breaking G at its CSEs. Each DFT is covered by means of tree parsing. In this way, the efficiency, local optimality, and easy retargetability of tree parsing can be exploited.

This means that we do not explicitly need to care about those parts of a DFT, which cannot be covered by SIMD instructions at all, because for those parts the standard tree parsing technique quickly generates optimal covers. In order to incorporate SIMD instructions and to generate alternative covers, we use a special formulation of the tree grammar for modeling the instruction set.

We will use the TI 6201 and the Philips Trimedia as concrete examples. Since the complete tree grammars (in OLIVE input syntax) in our formulation have a size of about 4,000 lines each, we will only highlight the SIMD-specific parts of the grammars. The other parts are rather straightforward and are easy to implement for users familiar with tree parsing.

4.1   TREE GRAMMAR FORMULATION
ARITHMETIC AND LOGICAL OPERATIONS

Consider a DFT segment representing the addition of two 16-bit ("short") values a and b, as shown in fig. 5.7 a). If there is a 32-bit addition instruction ADD as well as the SIMD instruction ADD2 (fig. 5.3), then we have three options of covering the addition.

- The addition is executed with full registers, i.e., the 32-bit ADD instruction is used (fig. 5.7 b). This is valid, because we can later store the least significant 16 result bits, while dropping the most significant 16 bits.

---

[6]For sake of simplicity, we assume that it is better to compute CSEs only once, instead of replicating the corresponding computations.

- The addition is covered by ADD2, while using the upper subregisters of the argument and destination registers (fig. 5.7 c). In this case, there has to be another DFG segment also representing a 16-bit addition, which is not yet known, but which will be executed on the lower subregisters within the *same* instance of the ADD2 instruction.

- Similar to case 2, we can use the lower part of ADD2 to cover the addition. In this case, the upper part needs to be determined later (fig. 5.7 d).

When covering one DFT at a time, it cannot be determined immediately, which option is valid and globally leads to the best result. This can only be decided when sufficient information about the entire DFG is available. We therefore keep all three options.

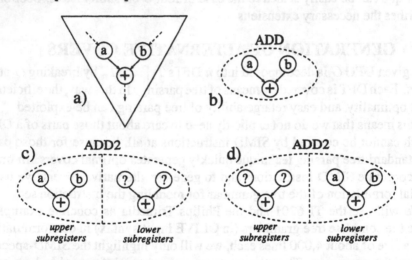

Figure 5.7.   Alternative covers of 16-bit addition

In the tree grammar, we need to distinguish full registers (denoted by the nonterminal reg) as well as upper and lower subregisters (denoted by the nonterminals reg_up and reg_lo). The usual ADD instruction is modeled by the grammar rule (in OLIVE syntax):

```
reg: PLUS(reg, reg)
```

while we use two different rules for modeling the upper and lower parts of the SIMD instruction ADD2:

```
reg_up: PLUS(reg_up, reg_up)
reg_lo: PLUS(reg_lo, reg_lo)
```

The latter two rules ensure that any addition, whose result is placed into the upper (lower) subregister of the full destination register also takes its arguments from the upper (lower) subregisters of the full argument registers.

Since the tree parser must not eliminate any of the three options, all three rules are assigned the same costs. Note that we do not count the costs of ADD2 twice: Rule costs are only used in the first phase of DFG covering in order to determine a set of alternative optimal DFT covers. As will be explained later, the rule costs are no longer considered in the second phase.

Other arithmetic operations as well as logical operations are modeled analogously. The concept of using dedicated nonterminals for subregisters also enables the modeling of the cross-wise multiply instructions of the TI 6201. The MPYLH instruction, for instance, is represented by the rule

```
reg: MULT(reg_lo, reg_hi)
```

CONSTANTS

A special treatment is required for instructions that load constants into registers. The Philips Trimedia, for instance, has an instruction "iimm" for loading a 32-bit constant (represented by the nonterminal const32) into a full register. The corresponding rule is

```
reg: const32
```

However, there is no instruction for directly loading a 16-bit constant into a subregister. If we have a DFG with two potentially parallel 16-bit additions of type "variable + constant", then a SIMD instruction could normally not be used. Therefore, instruction "iimm" has to be transformed into a SIMD instruction as well. This is achieved by introducing two additional rules:

```
reg_up: const16
reg_lo: const16
```

These rules allow to load one 16-bit constant (const16) into one subregister each by using the upper and lower parts of "iimm" simultaneously. Like for arithmetic SIMD instructions, we later have to ensure that none of these two rules are interpreted as "stand-alone" instructions.

LOADS AND STORES

Similar to arithmetic and logical operations, there are three options for loading a 16-bit "short" value from memory into a register: Either a full register (in which case a sign extension[7] is performed) or one of the two subregisters can be used. The corresponding grammar rules are:

```
reg:    LOAD_SHORT(addr)
```

---

[7]In reality, we have to distinguish between signed and unsigned values. This is omitted here for sake of brevity.

```
reg_up: LOAD_SHORT(addr)
reg_lo: LOAD_SHORT(addr)
```

Here, nonterminal addr denotes a memory address. Since the latter two rules do not represent valid instructions themselves, they always have to be used in pairs for a given DFG. In this case, a single 32-bit LOAD assembly instruction is emitted for both.

In order to allow this, an additional constraint has to be obeyed: If a value x is loaded into reg_up, and y is loaded into reg_lo, then these two values must be located in adjacent memory locations. For instance, the address difference of x and y must be 2 for a byte-addressable memory. Otherwise, a single 32-bit LOAD would not be able to access x and y simultaneously. Since this constraint cannot be checked locally for a DFT, the checking is postponed to the second phase.

Just like LOADs, we can also use STOREs as SIMD instructions for 16-bit data. If S is the grammar start symbol, then the corresponding set of rules is:

```
S: STORE_SHORT(addr,reg)
S: STORE_SHORT(addr,reg_up)
S: STORE_SHORT(addr,reg_lo)
```

The constraints for using the latter two are the same as for LOADs.

COMMON SUBEXPRESSIONS

SIMD instructions can also be used for covering CSEs in a DFG. As an example, consider fig. 5.8. Suppose, the DFTs T_1 and T_2 define 16-bit CSEs used in T_3, T_4, and T_5. These CSEs could be communicated either via two separate full registers or via the two subregisters of a single full register, in which case the roots of T_1 and T_2 were covered by an ADD2 instruction.

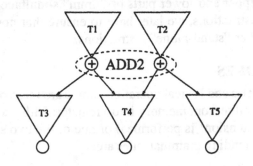

Figure 5.8.   Using SIMD instructions for common subexpressions

At the time of covering T_1 it is not known whether using ADD2 in conjunction with T_2 is possible. Therefore, also the alternative of using a full register must be kept. Similar to the modeling from chapter 3, we use dedicated terminals for representing definitions and uses of CSEs. For the definition of a 16-bit "short" CSE, we provide three alternative rules with identical cost values:

```
S: DEF_SHORT_CSE(reg)
S: DEF_SHORT_CSE(reg_up)
S: DEF_SHORT_CSE(reg_lo)
```

Likewise, there are three rules for using a 16-bit CSE:

```
reg:     USE_SHORT_CSE
reg_up:  USE_SHORT_CSE
reg_lo:  USE_SHORT_CSE
```

Similar to LOADs and STOREs, global consistency has to be preserved when applying the latter two rules. In fig. 5.8, for instance, assigning the results of T_1 and T_2 to reg_up and reg_lo, respectively, requires that all uses in T_3, T_4, and T_5 read the corresponding CSEs from the same subregisters to which they have been assigned.

4.2   ADAPTATION OF THE OLIVE CODE SELECTOR GENERATOR

For the implementation of DFT covering, we use the OLIVE tool, which has already been described in chapter 3. Using OLIVE makes it relatively easy to write a code selector for a certain target processor, which is very important from a practical viewpoint. However, OLIVE in its original form does not produce code selectors suitable for generating alternative DFT covers. Fortunately, it is possible to adapt the tool for SIMD instructions without affecting its functionality or the efficiency of the code selectors. This is briefly described in the following.

From a specification of a tree grammar G, OLIVE generates a code selector CS_G. Normally, the internal data structures of CS_G need not to be accessed by the user, since CS_G as a user interface comprises functions for traversing the derivation tree constructed for an input DFT T. Essentially, only the action function of the start symbol needs to be called in order to emit assembly code for T.

In our case, however, we also need to know the complete set of rules which *match* the inner nodes of T during parsing. We say that a rule R of the form $n_1 \rightarrow x(n_2, n_3)$ with nonterminals n_1, n_2, n_3 and a terminal x matches a node $v \in T$, labeled with operator x, if

- the rule R is the first rule in a derivation of the complete subtree rooted at v from n_1, and

- there is no other rule R' with the same property, which allows a derivation at lower total costs.

Note that in general there are multiple matching rules for a node v, which may have different nonterminals on their left hand sides (we will call these *target nonterminals*). The code selector CS_G generated by OLIVE keeps only one matching rule per target nonterminal. If there happen to be multiple rules for a target nonterminal n_1 matching v at the same total costs, then all but one of these rules are arbitrarily discarded.

This is sufficient if CS_G is applied in a usual tree parsing context, because only one (out of possibly many) optimal DFT covers has to be found. However, this causes problems when SIMD instructions must be exploited, which have the same target nonterminal. This is, for instance, the case for the multiply instructions of the TI C6201, which are modeled by five rules:

```
reg: MULT(reg, reg)          // pseudo-instruction
reg: MULT(reg_lo, reg_lo)    // MPY instruction
reg: MULT(reg_lo, reg_up)    // MPYLH instruction
reg: MULT(reg_up, reg_lo)    // MPYHL instruction
reg: MULT(reg_up, reg_up)    // MPYH instruction
```

The first one is used to implement 32×32-bit (i.e. "integer") multiplications, in which case a certain sequence of assembly instructions (instead of a single one) are generated. The latter four correspond to SIMD instructions, since they operate on subregisters. All five rules have equal costs[8] and they have reg as the target nonterminal. So four of them would normally be discarded during tree parsing.

In order to keep *alternative* matching rules for each DFT node v without giving up the numerous advantages of the OLIVE tool, we have modified OLIVE in such a way that its generated code selectors allow to retain multiple matching rules for each target nonterminal. More precisely, our modified OLIVE version generates code selectors CS_G, which at any node v annotate *all* matching rules during tree parsing.

Note that according to our above definition of *matching* rules, we do not actually keep *all* rules for a node v, using which the subtree rooted at v can be reduced. Those rules which do not allow a reduction at minimum costs are still discarded. In fact, this is the point where our technique potentially fails to find globally optimal solutions. We assume that the optimal DFG cover can be composed from the alternative optimal covers of its DFTs. This assumption is not necessarily fulfilled (even though it is difficult to find a practical case, for

---

[8]The first rule is assigned higher costs only for matching 32×32-bit multiplications, in which case a SIMD instruction cannot be used anyway.

which it is not), but it is required to ensure a reasonable runtime for the second phase in our approach.

4.3 TREE TRANSFORMATION RULES

An important point in code selection with SIMD instructions is the structure of the DFTs. In general, there are multiple alternative DFTs representing the same computation. Such alternatives, for instance, arise if *algebraic rules* (e.g. associativity or distributivity) are used to transform a DFT.

Using DFT transformations is common in compiler construction [Much97], so as to replace complex DFTs in the intermediate representation by simpler ones (for instance by detecting more opportunities for constant folding). Several compilers for embedded processors and hardware synthesis systems made use of transformation rules to generate alternative DFTs, so as to optimize code quality [Nowa87, Leup97] or hardware costs [NiPo91, LaMa97]. In these approaches, different alternatives are evaluated in detail, before the best one is finally selected.

Utilizing SIMD instruction also requires DFT transformations, because otherwise some opportunities for SIMD instructions cannot be detected at all. For instance, consider the C statement

$$y = c1 * x1 + c2 * x2 + c3 * x3 + c4 * x4;$$

The C syntax specification [KeRi88] prescribes a left-to-right parsing of the statement. Thus, it implies that the corresponding DFT in the intermediate representation is structured as if we had written:

$$y = (((c1 * x1 + c2 * x2) + c3 * x3) + c4 * x4);$$

However, in order to make use of the SIMD instruction "ifirii" from fig. 5.4 the following balanced structure were required:

$$y = (c1 * x1 + c2 * x2) + (c3 * x3 + c4 * x4);$$

By appropriately using parentheses in the source code, the programmer can enforce a certain DFT structure, but actually there is no need to burden the programmer with this. Since we do not require an extremely high compilation speed, we can generate a large number of alternatives for a given input DFT, each of which is covered. The best one is selected based on the cost metric returned by the tree parser.

In our approach, we use a fixed set of algebraic transformation rules to generate alternative DFTs. These include associativity rules (for detecting special SIMD instructions) as well as some further simple rules (for increasing code quality in general). These rules are stored in a library, and are recursively applied to all DFTs.

Exploiting commutativity, which is also extremely important for a good code quality, virtually comes for free: One can simply duplicate grammar rules representing commutative operations with different argument nonterminals, while exchanging the arguments. This increases the runtime of the code selector only by a small constant factor.

5.   COVER SELECTION

After generation of alternative covers, a set $M(v)$ of alternative matching rules has been determined for each DFG node v. For nodes v that qualify for covering by SIMD instructions (e.g. 16-bit additions or multiplications) $M(v)$ generally has a size larger than one. Therefore, in a second phase of code selection, we now determine a single rule from $M(v)$ for each v, such that a set of correctness constraints is met and that the use of SIMD instructions is globally maximized for a DFG.

Note that considering the rule costs is no longer required in this phase: By our definition of *matching* rules, all rules $R \in M(v)$ are guaranteed to have minimum, and therefore identical, costs. The only alternatives represented by $M(v)$ concern the selection of a certain target nonterminal for v.

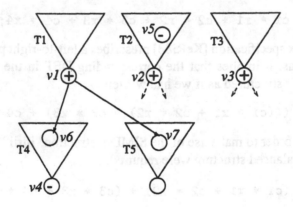

Figure 5.9.   Illustration of DFG covering constraints

5.1   COVERING CONSTRAINTS

We will first exemplify the constraints that have to be obeyed to obtain a valid DFG cover. Consider the situation in fig. 5.9. There are five DFTs, $T_1, \ldots, T_5$, the first three of which are supposed to have a 16-bit addition at their roots v_1, v_2, and v_3. The root v_4 of T_4 as well as some inner node v_5 of T_2 are labeled with a 16-bit subtraction. Finally, let v_6 and v_7 be the uses of a common subexpression computed by T_1. Informally, the following constraints have to be met:

Selection of a single rule: For each node v_i (i.e., $v_1, \ldots, v_7$, as well as all other DFG nodes) exactly one rule from $M(v_i)$ has to be selected. For v_1, v_2, and v_3, for instance, which are candidates for SIMD instructions, $M(v_i)$ contains the following rules:

```
(R1) reg:     PLUS(reg, reg)
(R2) reg_up:  PLUS(reg_up, reg_up)
(R3) reg_lo:  PLUS(reg_lo, reg_lo)
```

Consistency of target nonterminals: Let x and y denote the argument nodes of v_1 in DFT T_1. Since rules R_2 and R_3 have reg_up and reg_lo as their arguments, respectively, it is guaranteed (by the tree parser) that both $M(x)$ and $M(y)$ contain rules with the target nonterminals reg_up and reg_lo. The selection of a rule for v_1 implies the selection of rules for x and y. For instance, if R_2 is selected for v_1, then the rules selected for both x and y must have reg_up as their target nonterminal. Otherwise, the assignment to full, upper, or lower (sub)registers would be inconsistent.

Common subexpressions: The rule selected for CSE v_1 also has implications on the rules selected for its uses v_6 and v_7. If rule R_3 is selected, then the CSE is passed via a lower subregister, so that all uses have to read the value from a lower subregister as well.

Node pairing: A SIMD instruction, such as ADD2, always covers a *pair* of DFG nodes. In fig. 5.9, valid pairings are (v_1, v_2), (v_1, v_3), and (v_2, v_3). Therefore, for two of the nodes v_1, v_2, and v_3, one instance of an ADD2 instruction may be selected, while for the third node a usual 32-bit instruction must be chosen, since there is no "partner" node remaining. *Which* pairs are formed is generally decided during the optimization process.

Schedulability: Since we only *select* instructions but do not schedule them, it must be ensured that scheduling in a later phase is still possible. In fact, an unfavorable code selection can result in deadlock situations. In fig. 5.9, we might for instance cover v_1 and v_2 by a single ADD2 instruction. Due to data dependencies, v_5 has to be scheduled *before* v_2, while v_4 has to be scheduled *after* v_1. Even though v_4 and v_5 in principle could be covered by a SIMD instruction (in this case a "SUB2") as well because they are potentially parallel, this would prevent a successful scheduling: Since v_1 and v_2 are scheduled at the same time, the SUB2 instruction would need to be scheduled both before and after the ADD2, which is impossible. Therefore, covering v_4 and v_5 by SUB2 and covering v_1 and v_2 by ADD2 are *mutually exclusive* options.

Especially the required node pairing makes the selection of a DFG cover quite complex. Actually, there is a close relation to the *minimum set covering* problem, which is known to be NP-complete [GaJo79]: There is a *base set* of nodes coverable by SIMD instructions, and there is a set of possible pairings[9]. In order to maximize the utilization of SIMD instructions, a minimum subset of the possible pairings have to be selected, such that all nodes in the base set are covered.

In addition, a number of complex constraints have to be obeyed which makes it difficult to construct an algorithm for DFG covering. Therefore, we solve the problem by mapping it into an *Integer Linear Programming* (ILP) formulation and solving the ILP instead. ILP is the problem of maximizing a linear objective function of the form

$$f(x_1, \ldots, x_n) = c_1 \cdot x_1 + \ldots + c_n \cdot x_n$$

under the following system of constraints:

$$a_{11} \cdot x_1 + \ldots + a_{1n} \cdot x_n \leq b_1$$
$$\vdots \qquad\qquad \vdots$$
$$a_{m1} \cdot x_1 + \ldots + a_{mn} \cdot x_n \leq b_m$$

All values a_{ij}, c_i, and b_i are real constants, while $x_1, \ldots, x_n \in \mathbb{Z}$ are the *solution variables*. ILP is an NP-complete problem as well [GaJo79], but there exist *ILP solvers*, which make the exact solving of moderate size ILP problems possible in practice.

By translating some optimization problem into an ILP formulation, one can thus take advantage of the large amount of techniques that have been developed for ILP, which is a well-known standard optimization problem. This approach has also been taken in a number of previous works on code generation or hardware synthesis, e.g. [GeEl92, LMD94, WGHB94, LeMa95a, KaLa98]. An additional advantage of using an ILP formulation is that, once an "encoding" of the optimization problem at hand in the form of an ILP has been found, constraints on the solution space can easily be specified.

Of course, one critical point when using ILP are the runtime requirements. In our approach to code selection, much of the work is performed efficiently by tree parsing, based on the special tree grammar formulation described in the last section. Therefore, we can restrict our use of ILP to a relatively small optimization problem. As will be shown later, this makes the approach applicable for DFGs of realistic size.

5.2   ILP FORMULATION

Let $\{v_1, \ldots, v_n\}$ be the set of DFG nodes, and let R_j be a rule in the set $M(v_i)$ of all rules matching v_i. We define Boolean solution variables x_{ij} as

---

[9] If we include 8-bit SIMD instructions as described later, then also *quadruples* of nodes have to be considered during covering.

follows:

$$x_{ij} = \begin{cases} 1, & \text{if } R_j \text{ is selected for } v_i \\ 0, & \text{else} \end{cases}$$

The x_{ij} variables account for the detailed rule selection for DFG nodes and will also be used in our objective function.

Furthermore, we need auxiliary Boolean variables. In order to express whether a pair (v_i, v_j) will be covered by a single SIMD instruction, we introduce the notion of a *SIMD pair*. For a SIMD pair (v_i, v_j), the following constraints must hold:

- v_i and v_j must be potentially parallel, i.e., there must be no path in the DFG from v_i to v_j and vice versa.

- v_i and v_j must represent the same operator (arithmetic, logical, load, or store) and must be 16-bit operations. In case that v_i and v_j are constants, equality is not required, but both constants must fit into 16 bits each.

- $M(v_i)$ must contain a rule with target nonterminal reg_up, and $M(v_j)$ must contain a rule with target nonterminal reg_lo.

- If v_i and v_j are LOAD or STORE operations, whose arguments are the pointers (memory addresses) p_i and p_j, then the difference $p_j - p_i$ is equal to the number of memory words occupied by a 16-bit value. This constraint is required to enable the simultaneous access of v_i and v_j values with a single 32-bit LOAD/STORE (see also fig. 5.5).

If v_i and v_j are a SIMD pair, then they *can* (but not necessarily must) be covered by a single SIMD instruction. Note that these constraints need not to be considered at the time of solving the ILP, since they can be checked in advance. The latter constraint requires some effort in case that p_i and p_j are not constant. We have implemented a limited mechanism for symbolically computing address differences, such as found in case of array accesses in loops. This mechanism enables us to establish the difference of expressions like "$A[c_1 \cdot i + c_2]$" for constants c_1, c_2 and a loop variable i, even though "A" is a symbolic address. By means of *global data flow analysis* [ASU86], we can also track values of symbolic pointers across basic block boundaries.

If (v_i, v_j) is a SIMD pair, then we define the auxiliary variable y_{ij} by

$$y_{ij} = \begin{cases} 1, & \text{if } v_i \text{ and } v_j \text{ are covered by the same SIMD instruction instance} \\ 0, & \text{else} \end{cases}$$

Thus, $y_{ij} = 1$ means, that the result of v_i will be assigned to the upper subregister and the result of v_j will be assigned to the lower subregisters of the *same* full register. We now describe how the constraints exemplified above are generally expressed by means of the variables x_{ij} and y_{ij} in the ILP model.

Selection of a single rule: Each DFG node v_i must be covered by a single rule from $M(v_i)$. This is specified by the constraint

$$\forall v_i : \sum_{R_j \in M(v_i)} x_{ij} = 1$$

Consistency of target nonterminals: Let $R_j \in M(v_i)$ with $R_j = n_1 \rightarrow t(n_2, n_3)$ (unary operations are just a special case) for a terminal t and nonterminals n_1, n_2, n_3, and let v_l and v_r be the left and right arguments of v_i. Furthermore, for any node v, let $M^N(v) \subseteq M(v)$ denote the subset of rules matching v that have N as the target nonterminal. If $R_j = n_1 \rightarrow t(n_2, n_3)$ is selected for v_i, then the rules chosen for v_l and v_r must have the correct target nonterminals n_2 and n_3, respectively. This is enforced by the constraints

$$\forall v_i : \forall R_j \in M(v_i) : \quad x_{ij} \leq \sum_{R_k \in M^{n_2}(v_l)} x_{lk}$$

$$\forall v_i : \forall R_j \in M(v_i) : \quad x_{ij} \leq \sum_{R_k \in M^{n_3}(v_r)} x_{rk}$$

Common subexpressions: We had defined three rules for definitions of 16-bit CSEs, and three rules for the corresponding uses:

```
(R1) S: DEF_SHORT_CSE(reg)
(R2) S: DEF_SHORT_CSE(reg_up)
(R3) S: DEF_SHORT_CSE(reg_lo)

(R4) reg:    USE_SHORT_CSE
(R5) reg_up: USE_SHORT_CSE
(R6) reg_lo: USE_SHORT_CSE
```

If v_i is a CSE definition, and v_u is one of its uses, then it must be ensured that the location (either full register or upper or lower subregister) is identical for both. Since we attach the special nonterminals "DEF_SHORT_CSE" and "USE_SHORT_CSE" to CSE definitions and uses, which can only be parsed via the above six rules, it is clear that $M(v_i) = \{R_1, R_2, R_3\}$ and $M(v_u) = \{R_4, R_5, R_6\}$. Therefore, in the ILP model we can simply unify the corresponding x_{ij} variables as follows, so as to ensure consistency for CSEs:

$$\forall v_i, v_u : \quad x_{i1} = x_{u4}$$

$$\forall v_i, v_u : \quad x_{i2} = x_{u5}$$

$$\forall v_i, v_u : \quad x_{i3} = x_{u6}$$

If, for instance, CSE v_i is written to a full register (target nonterminal reg), then also all uses read from reg and vice versa.

Node pairing: Let P denote the set of all SIMD pairs according to the above definition, and let $(v_i, v_j) \in P$. Thus, v_i and v_j *could* be covered by the same SIMD instruction. We use auxiliary variable y_{ij} to link the rule selection for v_i and v_j. For any node v, let $M^{up}(v) \subset M(v)$ $(M^{lo}(v) \subset M(v))$ denote the subset of rules matching v, that work on upper (lower) subregisters. If a rule $R_k \in M^{up}(v_i)$ is selected for v_i, then v_j is covered by the "upper half" of a SIMD instruction. Thus there must exist a "partner" node v_j, such that $(v_i, v_j) \in P$, and v_j is covered by the "lower half" of the same SIMD instruction, which is expressed by setting $y_{ij} = 1$. A dual situation occurs, if some rule $R_k \in M^{lo}(v_i)$ is selected for v_i. We need two constraints to model this in the ILP:

$$\forall v_i : \qquad \sum_{R_k \in M^{up}(v_i)} x_{ik} = \sum_{j:(v_i,v_j)\in P} y_{ij}$$

$$\forall v_i : \qquad \sum_{R_k \in M^{lo}(v_i)} x_{ik} = \sum_{j:(v_j,v_i)\in P} y_{ji}$$

The first constraint ensures that for some j, y_{ij} will be set to 1, if v_i is covered by a rule in $M^{up}(v_i)$, and vice versa. The second constraint covers the dual case, that a rule in $M^{lo}(v_i)$ is selected. The left hand sides of the equations are always less or equal to 1. Thus, the same holds for the right hand sides, which in turn ensures that any node v_i will be covered by *at most* one SIMD instruction.

Schedulability: For any node v, let $X(v)$ denote the set of nodes that must be scheduled before v, and let $Y(v)$ denote the set of nodes that must be scheduled after v, according to the DFG dependencies. If the pair (v_i, v_j) is covered by a SIMD instruction, then no pair (v_k, v_l) in the set

$$Z_{ij} \quad := \quad P \quad \cap \quad (X(v_i) \times Y(v_j) \cup X(v_j) \times Y(v_i))$$

must be covered by a SIMD instruction in order to avoid scheduling deadlocks. This is ensured by the ILP constraint

$$\forall(v_i, v_j) \in P : \quad \forall(v_k, v_l) \in Z_{ij} : \quad y_{ij} + y_{kl} \leq 1$$

which enforces that *at most* one of (v_i, v_j) and (v_k, v_l) can be covered by a SIMD instruction.

Objective function: Our optimization goal is to make the maximum use of SIMD instructions for a DFG. This goal is achieved if, among all possible coverings, a maximum number of node pairs are covered by SIMD instructions. We maximize the number of selected rules representing SIMD instructions by using the following objective function:

$$f = \sum_{i=1}^{n} \sum_{R_j \in M^{up}(v_i) \cup M^{lo}(v_i)} x_{ij}$$

6.   EXTENSIONS FOR 8-BIT SIMD INSTRUCTIONS

In the above discussion, we have focused on the case of SIMD instructions that cover two 16-bit operations at a time. The techniques can easily be generalized to cover case of four 8-bit operations per SIMD instruction as well. For instance, this extension is required for the Philips Trimedia processor, which offers SIMD support both for the 16-bit and the 8-bit case.

Also for the TI C6201, we can sometimes exploit 8-bit SIMD instructions, even though there is no special hardware support. This concerns logical operations (e.g. performing four 8-bit ANDs in parallel with a 32-bit AND instruction) as well as LOADs and STORES.

The following extensions of the tree grammar formulation and the ILP model for cover selection are required:

- In the tree grammar, we introduce four additional nonterminals reg1, ..., reg4 to represent the four 8-bit subregisters of a full 32-bit register. For example, the SIMD instruction from fig. 5.4 is modeled by the rule

  ```
  reg: PLUS(PLUS(MULT(reg1,reg1), MULT(reg2,reg2)),
           PLUS(MULT(reg3,reg3), MULT(reg4,reg4)))
  ```

  Similar to the 16-bit case, we use four different rules to model the four operations performed by an 8-bit SIMD instruction, we provide rules that enable the loading of 8-bit constants into an 8-bit subregister, and we include register sharing for 8-bit common subexpressions. For any DFG node v that can potentially be covered by an 8-bit SIMD instruction, the tree parser then includes the 8-bit rules into the set $M(v)$ of alternative matching rules.

- In addition to the SIMD pairs defined for the 16-bit case, we use *SIMD quadruples* to capture node sets that qualify for covering with 8-bit SIMD instructions.

- In addition to the auxiliary variables y_{ij}, we use four-index Boolean variables y_{ijkl} to express whether a node quadruple (v_i, v_j, v_k, v_l) is covered by a single SIMD instruction. The constraints on these variables are formulated completely analogous to the case of 16-bit subregisters.

7. EXPERIMENTAL RESULTS

The code selection technique described above has been applied to two multimedia processors: the TI C6201 and the Philips Trimedia. We have used the modified OLIVE tool to implement code selectors for both processors. These code selectors construct DFG representations of basic blocs, compute the alternative covers, and automatically generate the concrete ILP problem instances required to perform the cover selection. Finally, an ILP solver[10] is invoked to solve the ILP, and assembly code is emitted in accordance with the binding of the x_{ij} solution variables.

The two code selectors have been applied to six C language routines from the area of digital signal and image processing: "vector add" is the example from fig. 5.5, "image compositing" is taken from [PWW97], and the remaining sources are from the DSPStone benchmark suite [ZVSM94]. All routines perform processing of 16-bit data.

The main goal of the experimentation was to verify that the technique is capable of exploiting SIMD instructions without the need for machine-dependent source code. Therefore, *identical* source codes in *plain* ANSI C have been used for both target processors. The consequences of exploiting SIMD instructions on code quality heavily depend on the detailed target instruction set.

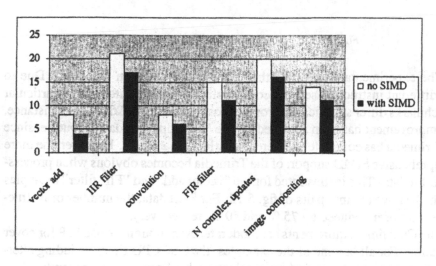

Figure 5.10. Code selection with SIMD instructions: Results for TI C6201

---

[10]In our experiments, we have used the public domain tool "lp_solve" from the TU Eindhoven [Eind00]. Also commercial solvers, such as IBM's OSL [IBM00], can be used, which typically require significantly lower runtime.

Fig. 5.10 shows the experimental results for the TI C6201. The columns show the number of instructions generated without (left) and with (right) using SIMD instructions. The SIMD support for this processor is relatively limited, and the maximum improvement (50 %) has been found for the "vector add" example.

An interesting result has been observed for the "N complex updates" routine, whose source code is shown in fig. 5.11. The use of SIMD instructions is not obvious here, because the TI C6201 has no "MULT2" instruction that computes two 16-bit products in parallel, and using a single 32-bit STORE for storing $D[i]$ and $D[i+1]$ is also impossible, since both values are in lower subregisters due to the occurrence of multiplications. However, our code selector was able to load 16-bit operand pairs (such as $A[i]$ and $A[i+1]$) with 32-bit LOADs, and to exploit the cross-wise subregister multiplication capabilities of the TI 6201. In total, this saves four LOAD instructions as compared to a solution without SIMD instructions.

```
short A[2*N], B[2*N], C[2*N], D[2*N]; // arrays of 16-bit values

for (i = 0 ; i < 2*N ; i += 2)
{ D[i]   = C[i] + A[i] * B[i] - A[i+1] * B[i+1];
  D[i+1] = C[i+1] + A[i+1] * B[i] + A[i] * B[i+1];
}
```

Figure 5.11.   Source code of "N complex updates"

The experimental results for the Trimedia are given in fig. 5.12. Due to the different instruction set, the code quality improvements for the particular benchmarks differ significantly from the results for the TI C6201. For instance, no improvement has been achieved for the "N complex updates" routine, since the Trimedia has completely different multiply capabilities. However, the more comprehensive SIMD support of the Trimedia becomes obvious when processing 8-bit data. This is illustrated for the "vector add" and "FIR filter" examples by the last two column pairs of fig. 5.12. For 8-bit data, the number of instructions have been reduced by 75 % and 50 %, respectively.

The CPU time requirements are moderate, even though we use ILP for cover selection. For all but one of the examples, the total CPU time, including alternative cover computation and cover selection, has been within 5 seconds. Only for the largest example ("vector add" on 8-bit data), for which the DFG contains 95 nodes, a significantly larger runtime (26.5 seconds) has been required, most of which was spent in cover selection. This indicates the practical limitations of our approach (even though we could still use faster ILP solvers). DFGs beyond a size of approximately 100 nodes should be split into smaller subgraphs in order to guarantee acceptable runtimes, possibly at the expense of lower code

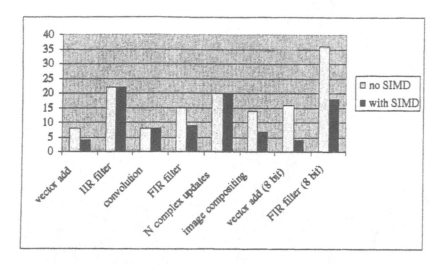

Figure 5.12. Code selection with SIMD instructions: Results for Trimedia TM1000

quality. A detailed result table, also comprising data on the number of ILP solution variables and constraints, can be found in appendix A (table A.6).

8. SUMMARY

SIMD instructions are a special feature of multimedia processors, which optimize the resource utilization when processing "short" data types. So far, however, compilers are not capable of automatically exploiting SIMD instructions without machine-dependent source code constructs. The technique presented in this chapter is a first approach to code generation with SIMD instructions for plain C source codes. For sake of simplicity, we neglected related issues such as memory alignment and automatic loop unrolling. However, it has been shown that code selection with SIMD instructions for real-life processors is possible even without machine-specific source code. In addition, our approach has two important advantages: When considering SIMD instructions already early during code generation, there is no need for special register allocation or scheduling techniques, because the mapping of values to subregisters is already completed after code selection. Furthermore, the proposed technique does not require any special DFG structure, such as symmetry, but it exploits SIMD instructions also in non-trivial cases.

Chapter 6

PERFORMANCE OPTIMIZATION WITH CONDITIONAL INSTRUCTIONS

The techniques presented in the previous chapters are strongly machine-dependent, since they are based on knowledge of the detailed data path architectures of embedded processors. In contrast, this chapter describes a largely machine-independent code optimization technique, which primarily works on the intermediate representation (IR) level of the compilation flow (fig. 6.1).

We assume that the target processor instruction set comprises *conditional instructions*, which we exploit for performance optimization. Conditional instructions allow for branch-less implementation of if-then-else statements, which are normally mapped to conditional jump constructs at the assembly level. This offers a number of advantages especially for deeply pipelined and highly parallel processors. However, using conditional instructions instead of conditional jumps is not always favorable. The decision which of these two alternative implementation schemes results in higher performance is difficult, particularly for

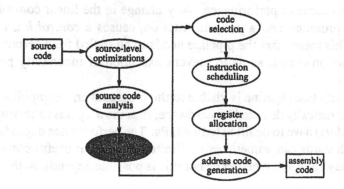

Figure 6.1.   IR optimizations in the compilation flow

127

nested if-then-else statements. We present a dynamic programming algorithm which globally optimizes such statements, based on estimated block execution times, by selecting the implementation schemes for if-then-else statements at all nesting levels simultaneously.

1.   CONDITIONAL INSTRUCTIONS

Compilers usually make use of conditional jump instructions in order to generate code for if-then-else (ITE) statements. This is exemplified in fig. 6.2, where the left part shows a C code fragment and the right part shows the corresponding assembly code (in a pseudo-syntax).

```
if (a > b)              CMP a,b        // compare a and b
{                       JG _L1         // jump if greater
  x = a + b;            SUB a,b,x      // else part
  y = x + 1;            SUB x,2,y
}                       JMP _L2        // unconditional jump
else              _L1:  ADD a,b,x      // then part
{ x = a - b;            ADD x,1,y
  y = x - 2;      _L2:                 // control flow joins here
}
```

Figure 6.2.   Implementation of an if-then-else statement with a conditional jump

Since in the context of embedded systems it is common that real-time constraints have to be met, we are interested in the *worst-case execution time* W of the code, measured in instruction cycles. This time is given by $W = \max(W_T, W_E)$, where W_T and W_E denote the execution times of the ITE statement in case the then or the else part gets executed, respectively. If we assume that each instruction requires one cycle, then we have $W = \max(4, 5) = 5$.

However, this assumption is not fulfilled, if the processor uses instruction pipelining to increase performance. Any change in the linear control flow of a machine program, i.e., a jump instruction, causes a *control hazard* in the pipeline. This means that the pipeline needs to be stalled for a certain number J of instruction cycles, so as to prevent execution of incorrectly prefetched instructions.

If the instruction pipeline is visible to the programmer, i.e. pipeline hazards are not automatically detected by hardware, then the J cycles following a jump (the *delay slots*) have to be filled with NOPs. The performance degradation due to control hazards can sometimes be limited by placing useful computations into the delay slots, but whether or not this is possible depends on the concrete program.

In the following, we will call J the *jump penalty*, i.e. the number of instruction cycles potentially lost by executing a jump instruction. The jump penalty can be very significant for deeply pipelined processors. The Philips Trimedia TM1000, for instance, has a jump penalty of 3, while for the TI C6201 the value is 5. For conditional jumps, the jump penalty does not depend on whether or not the jump is actually taken. Thus, for our example from fig. 6.2, the worst-case execution time on a TI C6201 is given as $W = \max(1+6+1+1, 1+6+1+1+6) = 15$ cycles, which is surprisingly high for such a simple program.

In order to avoid low performance in case of control-intensive programs[1], processor architects have introduced the concept of *conditional instructions* (also called *predicated execution*). The main idea is to attach a Boolean *guard* condition c to each regular instruction I. At program runtime, the value of c (typically stored in a register) at the moment when the control flow reaches I, decides whether or not I gets executed: If c is true, then I behaves like a usual instruction, otherwise it behaves like a NOP. A conditional jump instruction is nothing but a special case of a conditional instruction, where c is the jump condition and I is an unconditional jump.

As already observed in [AKPW83], replacing conditional jumps by conditional instructions essentially means to replace control dependencies by data dependencies. This process is also called *if-conversion*.

It is important to note that some conditional instruction (c, I) always consumes the same amount of resources, independent of the value of c. Thus, c can be viewed as being connected to the enable input of the destination register of I, so that c only decides whether or not the result computed by I will be stored. If not, then I does not change the processor state so that it effectively becomes a NOP.

There are different forms of conditional instructions in real-life processors. In the ARM RISC processor, as well as in the Analog Devices ADSP-210x DSP, conditions must be stored in condition code or flag registers. Since many instructions overwrite flags as a side effect, this is a rather volatile type of conditional instructions, i.e., a certain condition value can usually not "survive" for a large number of cycles. Thus, conditions stored in flags can only be used for conditionally executing very small pieces of code.

Another form of conditional instructions is given by conditional *skipping* of instructions. For instance, this is implemented in the HT48100 microcontroller and the TI C5x DSP. These processors have special instructions that allow to skip the execution of the following one or two instructions, dependent on some condition. Like in the above case, only small pieces of conditional code can be implemented in this way.

A more powerful and general form of conditional instructions is implemented in the TI C6201 and the Trimedia processors. There, conditions can be stored

---

[1]An example of highly control-intensive code is the ADPCM speech coding algorithm from the DSPStone benchmark suite [ZVSM94].

in general-purpose registers, which means that conditions may have (almost) arbitrarily long lifetimes. Also the next generation of 64-bit Intel processors will be equipped with this feature [Hwu98]. If conditions are stored in regular data registers, then large pieces of code can be executed conditionally, which in turn allows to avoid a large number of time-consuming jumps.

The reduction of the number of jumps implies another benefit, which is extremely important particularly for VLIW processors, since these show a large amount of instruction-level parallelism. As the use of conditional instructions effectively results in larger basic blocks, there are more opportunities of exploiting instruction-level parallelism by careful instruction scheduling, so as to increase performance.

2.   CONDITIONAL INSTRUCTIONS VERSUS CONDITIONAL JUMPS

The use of conditional instructions for efficient implementation of ITE statements is illustrated in fig. 6.3. Instead of using a single ITE statement, one can attach the condition c, as well as the negated condition, to all statements inside the then and else parts, respectively. In the assembly code in the right column of fig. 6.3, conditional instructions are denoted by attaching "[c]" or "[!c]" to the regular instructions. The left column shows how this could be written correspondingly at the C level, even though programmers would hardly do so for sake of readability.

```
c = a > b;                      CMPGT a,b,c   // compare for greater
if (c)  x = a + b;         [c]  ADD a,b,x     // then part
if (c)  y = x + 1;         [c]  ADD x,1,y
if (!c) x = a - b;         [!c] SUB a,b,x     // else part
if (!c) y = x - 2;         [!c] SUB x,2,y
```

Figure 6.3.   Implementation of an if-then-else statement with conditional instructions

Obviously, this code is equivalent to the one from fig. 6.2 but it requires no jumps. On a VLIW processor, such as the TI C6201, the then and else parts, which still are mutually exclusive, can be parallelized. Since there is only one basic block, the execution time is independent of the condition c and amounts to only 3 cycles, a reduction by 80 % as compared to the conditional jump implementation.

However, also the use of conditional instructions has its limits. As we have shown, conditional instructions may result in a performance gain for ITE statements, but this is not guaranteed. Consider an ITE statement S with a condition c and basic blocks B_T and B_E in its then and else parts, and let J be the jump penalty. According to the ITE implementation scheme with conditional jumps (see fig. 6.2), one jump is executed (but not necessarily taken) if c is true, and two jumps are executed in case that c is false. Thus, the worst case execution time of S is given by

$$W(S) = \max(T(B_T) + J, T(B_E) + 2J)$$

where $T(B_T)$ and $T(B_E)$ are the constant execution times of B_T and B_E. If S is implemented with conditional instructions, then the blocks B_T and B_E are virtually *merged* (denoted by a "$\circ$" symbol) to form a single "large" block. In this case, the worst case execution time is constant and is equal to $W(S) = T(B_T \circ B_E)$. The execution of a merged block can be viewed as multiple threads of control being executed in parallel, only one of which gets *effectively* executed according to the current condition values.

In total, the use of conditional instructions results in a lower worst case execution time, exactly if

$$T(B_T \circ B_E) < \max(T(B_T) + J, T(B_E) + 2J)$$

Due to the generally large number of functional units available in a VLIW processor, it may well be the case that $T(B_T \circ B_E)$ is equal to or only slightly larger than $\max(T(B_T), T(B_E))$, because the instructions from the smaller of the two blocks might fit into instruction slots not occupied by the other one. This is illustrated in fig. 6.4 a) for a machine with two instruction slots.

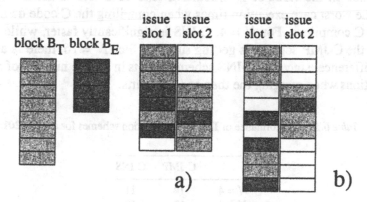

Figure 6.4. Parallelization of two conditional basic blocks

However, as shown in fig. 6.4 b), it may also happen that due to resource contentions between instructions in B_T and B_E the merger of both blocks requires a schedule length significantly larger than $\max(T(B_T), T(B_E))$. Whenever this difference outweighs the jump penalty (either J or $2J$), then using conditional instructions would increase the worst case execution time.

We use a small example to illustrate the threshold between conditional instructions and conditional jumps in practice. In the following we use the abbreviations C-JMP and C-INS to denote an ITE implementation by conditional

```
          int A[N+1],B[N+1];

          void f1(int c)
          {
            if (c)
            {
              A[1] = 1;
              ...
              A[N] = N;
            }
            else
            {
              B[1] = 1;
              ...
              B[N] = N;
            }
          }
```

Figure 6.5.   Example C code

jumps or conditional instructions, respectively. For the C code from fig. 6.5
it depends on the value of N whether C-JMP or C-INS is better. Table 6.1
gives the worst case execution times when compiling the C code using the TI
C6201 C compiler[2]. For $N = 4$, C-INS is significantly faster, while from N
= 8 on the C-JMP scheme is getting superior. For $N = 16$, there is already a
huge difference, since the C-INS scheme results in a large number of resource
contentions when merging the then and else parts.

Table 6.1.   Performance of ITE implementation schemes for a TI C6201

	C-JMP	C-INS
$N = 4$	14	11
$N = 8$	17	18
$N = 16$	25	62

In order to determine the best ITE implementation scheme before code generation it is obviously necessary to estimate the execution times of basic blocks
as well as mergers of basic blocks in case that C-INS is used.

---

[2]We have used a specific programming style as in fig. 6.3 to enforce the generation of conditional code by
the C compiler.

3. NESTED ITE STATEMENTS AND ITE TREES

So far we have considered ITE statements, where the then and else parts B_T and B_E are basic blocks. However, in general B_T and B_E may in turn comprise ITE statements, i.e., we have to deal with nested ITE statements. As we have shown, using C-INS for merging B_T and B_E results in multiple, mutually exclusive, threads of control being executed in parallel. Since the number of these parallel threads is only limited by the number of registers available for storing conditions, also nested ITE statements can be executed in the form of a single merged basic block, that consists of conditional instructions.

We represent a structure of nested ITE statements by means of an *ITE tree*. An ITE tree $T = (R, B_T, B_E)$ consists of a root R (a Boolean expression) and two *ITE blocks* B_T and B_E, called the *then block* and the *else block* of T, respectively. An ITE block is a sequence $(s_1, \ldots, s_n)$, such that each s_i is either a *simple statement*, i.e., a three-address code statement as defined in chapter 3, or s_i is in turn an ITE tree.

For sake of simplicity we assume that all simple statements inside any ITE block are assignments, and that there are no further control flow changes by means of function calls or loops inside an ITE block. Fig. 6.6 illustrates how the ITE tree data structure is used to represent a nested ITE statement.

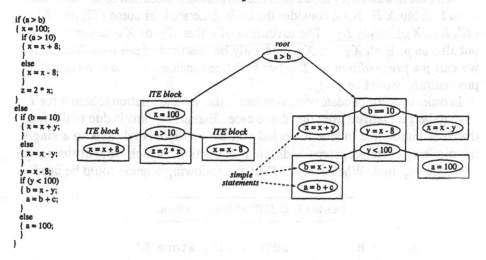

```
if (a > b)
{ x = 100;
  if (a > 10)
  { x = x + 8;
  }
  else
  { x = x - 8;
  }
  z = 2 * x;
}
else
{ if (b == 10)
  { x = x + y;
  }
  else
  { x = x - y;
  }
  y = x - 8;
  if (y < 100)
  { b = x - y;
    a = b + c;
  }
  else
  { a = 100;
  }
}
```

Figure 6.6. Nested ITE statement in C code and its ITE tree representation

Suppose the result of the root of an ITE tree $T = (R, B_T, B_E)$ is stored in a register p. If T is implemented by C-JMP, then machine code for T would be generated according to the following scheme.

Scheme 1: C-JMP without precondition

```
      p = R        // store condition R
  [p] goto L1      // goto then block
      B_E          // execute else block
      goto L2      // skip then block
L1:   B_T          // execute then block
L2:                // control flow joins
```

Conversely, when using C-INS the implementation scheme would be as follows.

Scheme 2: C-INS without precondition

```
      p = R        // store condition R
      q = !p       // compute negated condition
  [p] B_T          // conditionally execute then block
  [q] B_E          // conditionally execute else block
```

Here, the notation "[c] B" denotes the conditional execution of all statements in an ITE block B. Now, consider the code generated for some ITE tree $T' = (R', X_T, X_E)$ inside B_T. The execution of either X_T or X_E depends on R' but also on p: Both X_T and X_E must only be executed, if p is true. Therefore, we call p a *precondition* for T'. For ITE trees inside B_E, the corresponding precondition would be $q = !p$.

In order to accommodate preconditions[3], the implementation schemes for T' have to be generalized from the above case. Essentially, this is due to the fact, that it is impossible to simultaneously attach multiple conditions to a single instruction. Therefore, some additional instructions, which we call the *setup code*, are required. When using C-JMP, the following scheme could be used.

Scheme 3: C-JMP with precondition

```
  [p] c = R'       // conditionally store R'
      q = !p       // compute negated precondition
  [q] c = 0        // reset c if precondition is false
  [c] goto L1      // goto then block
```

---

[3]Here, we assume that the target processor does not *directly* support negated conditions, but that negated conditions have to be explicitly computed. For instance, this is the case for the Philips Trimedia. If negated conditions are directly supported (e.g. in the TI C6201), then the implementation schemes presented here look slightly different. The same holds for ITE statements without an else part, i.e. "pure" if-statements. We omit the discussion of these special cases here for sake of brevity.

```
        [p] X_E      // conditionally execute else block
            goto L2  // skip then block
    L1:     X_T      // execute then block
    L2:              // control flow joins
```

The correctness of this scheme can be shown as follows. First, the result of the condition represented by R' is stored in some register c. Attaching the precondition p to this statement is actually only required if the evaluation of R' might cause undesired side effects. Next, the negated precondition q is computed, and c is set to zero (i.e. logically false), if the precondition is false as well. Then, a jump to X_T takes place, if c is true. This is the case, if and only if both the precondition p and the inner condition R' are true, which obviously is the correct condition for executing X_T.

If the jump to X_T is not taken, then X_E has to be executed, if and only if p is true and R' is false. If R' is false, then c is guaranteed to be false as well, so that the statement "[p] X_E" cannot be reached at all. Therefore, it is sufficient to execute X_E only under the precondition p. Next, a jump to the join label "L2" is executed. The code for X_T needs no further conditions, since when reaching the first statement in X_T all required conditions have already been checked.

If T' is implemented by C-INS, then the following implementation scheme is required.

Scheme 4: C-INS with precondition

```
    [p] c = R'    // conditionally store R'
        d = !c    // compute negated condition
        q = !p    // compute negated precondition
    [q] c = 0     // reset c if precondition is false
    [q] d = 0     // reset d if precondition is false
    [c] X_T       // conditionally execute then block
    [d] X_E       // conditionally execute else block
```

Like in the case of C-JMP, the condition R' is evaluated into a register c, and the negated precondition q is computed. Also the negated inner condition $d = !c$ is required. If the precondition is false, then neither X_T nor X_E must be executed. This is achieved by setting c and d to zero, if q is true. Now, block X_T is executed if c is true, i.e., both p and R' are true. Likewise, X_E is executed, if d is true, which implies that R' is false and p is true.

4.   PROBLEM DEFINITION

We now have four different implementation schemes for an ITE tree T: either C-JMP or C-INS, each either with or without a precondition. Each of these four schemes will generally result in a different execution time for T. Our optimization goal is the following: Given an "outermost" ITE tree T, i.e., T is not contained in another ITE tree[4], we would like to select either C-JMP or C-INS for T and all its inner ITE trees, such that the worst case execution time of T is minimized. Fig. 6.7 gives an example showing that such an optimization is relevant for real-life processors.

```
          C source code

          if (x > 10)
          { if (y > 13) z = x + y; else z = x - y;
          }
          else
          { z = x + y;
          }


          C-JMP/C-CMP              C-JMP/C-INS

          CMPGT x,10,p            CMPGT x,10,p
     [p]  JMP L1             [p]  JMP L1
          MUL x,y,z               MUL x,y,z
          JMP L2                  JMP L2
     L1:  CMPGT y,13,c       L1:  CMPGT y,13,c
     [c]  JMP L3                  NOT c,d
          SUB x,y,z          [c]  ADD x,y,z
          JMP L2             [d]  SUB x,y,z
     L3:  ADD x,y,z          L2:
     L2:


          C-INS/C-JMP             C-INS/C-INS

          CMPGT x,10,p           CMPGT x,10,p
          NOT p,q               NOT p,q
     [p]  CMPGT y,13,c       [p]  CMPGT y,13,c
     [q]  MOV 0,c                 NOT c,d
     [c]  JMP L1            [q]  MOV 0,c
     [p]  SUB x,y,z         [q]  MOV 0,d
          JMP L2            [c]  ADD x,y,z
     L1:  ADD x,y,z         [d]  SUB x,y,z
     L2: [q] MUL x,y,z
```

Figure 6.7. Illustration of the C-JMP and C-INS ITE implementation schemes: Four alternative assembly programs for a piece of C code are shown, which reflect the four possible combinations of using C-JMP and C-INS for two nested ITE statements. When applying the TI 6201 assembly optimizer to these codes, the worst scheme turns out to be C-INS/C-JMP with a worst case execution time of 18 cycles. Both C-JMP/C-INS and C-JMP/C-CMP require 12 cycles. The best solution in this example is C-INS/C-INS with only 8 cycles. The TI 6201 C compiler selects the C-JMP/C-INS scheme when applying it to the C source code. In contrast, the optimization algorithm presented in the following correctly predicts that C-INS/C-INS will be faster.

The main difficulty in this optimization problem is that if T comprises a total of n ITE trees, then there are 2^n possible combinations of C-JMP and C-INS, each of which may be the best solution. Since the globally best implementation

---

[4]At the C level, this corresponds to an ITE statement in the outermost scope within a function.

scheme for each ITE tree inside T generally depends on all other ITE trees, the selection of C-JMP or C-INS cannot be made locally. This is due to a twofold effect:

Bottom-up dependence: The choice of C-JMP or C-INS for an ITE tree $T = (R, B_T, B_E)$ obviously depends on the execution times of any inner ITE tree T' contained in B_T or B_E. In turn, the execution time of T' depends on whether C-JMP or C-INS are chosen for T'.

Top-down dependence: Let T be an ITE tree, and let T' be its "father" in the tree structure. Since the implementation schemes 1 and 3, as well as 2 and 4, are pairwise different due to the need for setup code, the execution time for T obviously depends on whether or not T has a precondition. A possible precondition is passed downwards from T' to T. According to the implementation schemes 1 to 4, there are exactly three cases in which T gets a precondition from T':

1. T' is implemented by scheme 2.
2. T' is implemented by scheme 3, and T lies in the else block of T'.
3. T' is implemented by scheme 4.

Thus, the selection of C-JMP or C-INS for T also depends on whether its father T' is implemented by C-JMP or C-INS.

Due to these bidirectional dependences, the best solution cannot be determined by a single bottom-up or top-down pass in the ITE tree. Fortunately, the mutual dependence between local solutions in an ITE tree is limited. After a review of related work, we will show how the tree structure of ITE statements can be exploited to compute the global optimum efficiently.

5. RELATED WORK

An effective compiler for VLIW processors with support for conditional instructions should aim at a good balancing between C-JMP and C-INS so as to maximize performance of control-intensive code. This has already been stressed in [Hwu98]: *With the adoption of advanced full predication support [...] predicated execution may become one of the most significant advances in the history of computer architectures and compiler design.*

Nevertheless, the number of compilation techniques exploiting conditional instructions is still low. Gupta et al. [GBF97] proposed to exploit conditional instructions for improving the performance of *partial dead code elimination* (PDE). PDE [KRS94] aims at detecting computations whose result is not needed on some of the possible control flow paths[5]. Moving such computations down-

---
[5] A dual optimization is *partial redundancy elimination* [KRS92].

wards in the control flow graph of a function to those points where they are actually used can increase performance. In [GBF97] conditional instructions are used to obtain a higher degree of freedom for moving partially dead statements. Since this technique works on a control flow graph representation of functions, it does not actually select between C-JMP or C-INS for ITE statements. Instead, conditions are selectively copied to serve as predicates for statements that have been moved.

The technique described in [MLC+92] is based on the formation of *hyperblocks*. A hyperblock is an extended basic block, in which multiple threads of conditional code are executed concurrently. In our notation, this corresponds to the merging of then end else parts of ITE statements by attaching the appropriate conditions to instructions. Thus, placing two basic blocks into the same hyperblock means to prefer the C-INS implementation scheme over C-JMP. The decision whether a basic block is included in a hyperblock is based on several heuristics. These take into account the block execution frequency as well as the block size. However, the mutual dependencies between different blocks at not considered, as the formation of hyperblocks is performed on a block-by-block basis.

Several deficiencies of that approach have been eliminated in [AHM97]. The most important improvement is that the final formation of hyperblocks is guided by detailed schedulability information. This is achieved by partially integrating if-conversion into the scheduling process. More specifically, the instruction scheduler is allowed to revise decisions on hyperblock formation that have been made earlier. This results in an increased accuracy in predicting the performance effects of if-conversion, which eventually results in better schedules.

One main difference of that work to our approach presented in the following section is that we explicitly consider the overhead due to possibly required additional instructions in the form of setup code for preconditions. That is, we take both the bottom-up and top-down dependencies mentioned above into account. In contrast to [AHM97], we strictly separate the selection of C-JMP and C-INS from the estimation of basic block execution times. Thus, we have to rely on accurate estimations. However, given an accurate estimation, our algorithm guarantees an optimal solution for a nested ITE statement with respect to its worst case execution time. From a practical viewpoint, our approach has the advantage that we can decouple if-conversion from the actual machine code generation phase. This guarantees a certain degree of machine-independence and permits the reuse of existing tools.

6. OPTIMIZED ITE IMPLEMENTATION BY DYNAMIC PROGRAMMING

The optimization algorithm presented here is based on *dynamic programming* (DP). The DP method can be applied to decomposable optimization problems, where optimal solutions can be quickly computed from combinations of optimal solutions for "smaller" subproblems. Lengauer [Leng90] gives different examples of using DP in VLSI layout synthesis. Another example is the tree parsing method for code selection discussed in chapter 3. Like in tree parsing, we exploit the tree structure of ITE statements in order to make DP applicable.

Our DP algorithm makes two passes over a given ITE tree T. The first one is a bottom-up pass, during which *cost tables* are attached to any ITE tree root inside T as well as to the root of T itself. The cost tables contain one entry for each of the above four implementation schemes, which represents the (estimated) worst case execution time. Whether C-JMP or C-INS will finally be selected cannot be decided during the first pass, and also the presence of a precondition for T is not yet known, since this depends on the scheme selected for the father of T. However, the cost table for any root can be computed only based on cost table information from its ITE subtrees.

The second pass traverses T in top-down direction. Starting from the root R of T, either C-JMP or C-INS are selected in such a way, that the worst case execution time of the complete ITE tree rooted at R is minimized, based on estimated execution times of the ITE blocks. The decision made for R also creates the information required to select either C-JMP or C-INS for its ITE subtrees. This traversal is recursively continued until all leaves have been processed. The two phases are described in more detail in the following.

6.1 COST TABLE COMPUTATION

The cost table computation takes two components into account in order to determine a cost value for the root R of some ITE tree $T = (R, B_T, B_E)$. These are the setup costs, as well as the cost values previously computed for the ITE blocks B_T and B_E.

SETUP COSTS

As shown earlier, each of the four ITE implementation schemes is associated with a certain setup cost. These cost reflect the extra instructions that are required for computing conditions and accommodating possible preconditions. We do not count the instructions required for computing an ITE condition itself, because the corresponding costs are identical for all schemes and we are only interested in their cost differences.

For scheme 1, there is no setup code at all, while for scheme 2 there is one extra instruction required to compute the negated condition $q = !p$. There is more setup code in case that preconditions have to be taken into account. Scheme 3 requires two extra instructions, while scheme 4 needs four setup instructions before the actual processing of then end else blocks can take place. The setup costs are summarized in table 6.2.

Table 6.2.   Setup cost table for ITE implementation schemes

	C-JMP	C-INS
without precondition	scheme 1: 0	scheme 2: 1
with precondition	scheme 3: 2	scheme 4: 4

We can easily generalize the setup cost table if we would like to include the case of ITE statements without an else part, as well as processors with or without support for negated conditions. Like in the above discussion, we omit these special cases here. For our purposes it is mainly important that the setup cost table only contains *constant* values. Therefore, a simple table lookup is sufficient to retrieve the setup costs of a certain scheme. In the following, we denote the four setup cost entries by

$$S_{C-JMP}, \quad S_{C-INS} \quad \text{(schemes without precondition)}$$

and

$$S^P_{C-JMP}, \quad S^P_{C-INS} \quad \text{(schemes with precondition)}$$

ITE BLOCK COSTS

Besides the setup costs, we need to determine cost values for ITE blocks. According to the definition, an ITE block B is a sequence of statements $(s_1, \ldots, s_n)$. The costs of B are defined as the sum of the costs over all s_i. We need to distinguish two statement cost values, $C(s_i)$ and $C^P(s_i)$, dependent on whether or not s_i is executed under a precondition. Likewise, we use the notations

$$C(B) = \sum_{i=1}^{n} C(s_i)$$

$$C^P(B) = \sum_{i=1}^{n} C^P(s_i)$$

to denote the costs of the complete block without and with a precondition. If s_i is a simple statement, then we set

$$C(s_i) = C^P(s_i) = 1$$

The exact time to execute a simple statement s_i is of less importance here, because we are just interested in *comparing* execution times with and without precondition. $C(s_i)$ and $C^P(s_i)$ are set equal, since the execution time of any machine instruction I is independent of whether or not I is conditional. Thus, execution of s_i will take identical time in both cases.

If s_i is an ITE tree (R, B_T, B_E), then its cost actually depends on the possible presence of a precondition. Since we compute cost tables bottom-up, we already know the cost values

$$C(B_T), \quad C(B_E), \quad C^P(B_T), \quad C^P(B_E)$$

which denote the execution times of B_T and B_E without and with a precondition, respectively. The cost table for s_i is now computed as follows, with one entry for each possible implementation scheme:

Scheme 1: The setup costs S_{C-JMP} have to be used, and neither B_T nor B_E get a precondition. The total worst case execution time is estimated in accordance with our earlier analysis as:

$$C_{C-JMP}(s_i) = S_{C-JMP} + \max(C(B_T) + J, C(B_E) + 2J)$$

Scheme 2: The setup costs S_{C-INS} have to be used, and both B_T and B_E get a precondition. The total execution time is given by

$$C_{C-INS}(s_i) = S_{C-INS} + C^P(B_T \circ B_E)$$

The term $C^P(B_T \circ B_E)$ denotes an estimation of the execution time of the block resulting from merging B_T and B_E with mutually exclusive preconditions.

Scheme 3: The setup costs S_{C-JMP}^P have to be used, and only B_E gets a precondition. Thus, we have

$$C_{C-JMP}^P(s_i) = S_{C-JMP}^P + \max(C(B_T) + J, C^P(B_E) + 2J)$$

Scheme 4: The setup costs S_{C-INS}^P have to be used, and B_T and B_E are merged like for scheme 2:

$$C_{C-INS}^P(s_i) = S_{C-INS}^P + C^P(B_T \circ B_E)$$

Since we do not rely on schedulability information of statements, we use a relatively simple heuristic estimation of $C^P(B_T \circ B_E)$. As we are dealing with VLIW processors with a large amount of instruction-level parallelism, the time needed to execute the merger of B_T and B_E will typically be smaller than the time for executing one block after another, i.e.

$$C^P(B_T \circ B_E) < C^P(B_T) + C^P(B_E)$$

We use a machine-dependent parameter K, which has to be determined empirically, in order to reflect the number of functional units in a concrete target processor.

Without loss of generality, let B_T be the block with the higher execution time. If $C^P(B_T) \gg C^P(B_E)$ then it is likely that the instructions from B_E will almost completely fit into instruction slots not occupied by B_T. In contrast, if $C^P(B_T) \approx C^P(B_E)$, the possible parallelization is presumably more limited. We thus heuristically assume that when scheduling B_T a number $K \cdot C^P(B_T)$ of instruction slots remain free for scheduling instructions from B_E.

In addition, we incorporate the *depth* D of the ITE statement which B_T and B_E belong to in the global ITE tree. Innermost ITE statements have a depth of $D := 1$, and D grows by 1 for each higher level in the ITE tree. The ITE blocks tend to get larger with an increasing D, which generally implies a larger number of resource conflicts between B_T and B_E. Thus, a high depth tends to reduce the parallelization effect.

Taking these observations into account, we estimate $C(B_T \circ B_E)$ as follows:

$$C_{\max} = \max(C^P(B_T), C^P(B_E))$$
$$C_{\min} = \min(C^P(B_T), C^P(B_E))$$
$$c = \min(\tfrac{K}{D} \cdot C_{\max}, C_{\min})$$
$$C^P(B_T \circ B_E) = C^P(B_T) + C^P(B_E) - c$$

The "min" in the definition of the correction factor c ensures that $C^P(B_T \circ B_E)$ cannot be estimated less than $C_{\max}$, which would be impossible.

The values $C_{C-JMP}(s_i)$, $C_{C-INS}(s_i)$, $C^P_{C-JMP}(s_i)$, and $C^P_{C-INS}(s_i)$ constitute the cost table for the root of the ITE tree s_i. In case that s_i is contained in an ITE block B (i.e., s_i represents some inner ITE statement), we still need to define the cost values $C(s_i)$ and $C^P(s_i)$, which are needed to determine the total costs of B at the next higher level dependent on the presence or absence of a precondition for s_i. Since we aim at a minimum execution time, these values are defined by

$$C(s_i) \quad = \quad \min(C_{C-JMP}(s_i), C_{C-INS}(s_i))$$
$$C^P(s_i) \quad = \quad \min(C^P_{C-JMP}(s_i), C^P_{C-INS}(s_i))$$

6.2   SELECTION OF C-JMP OR C-INS

After the bottom-up pass, cost tables are available at all ITE tree roots. These reflect the costs of using the four possible implementation schemes for the corresponding ITE statement. Fig. 6.3 shows the structure of each cost table.

Table 6.3. Cost table structure for an ITE tree root R

	C-JMP	C-INS
without precondition	$C_{C-JMP}(R)$	$C_{C-INS}(R)$
with precondition	$C^P_{C-JMP}(R)$	$C^P_{C-INS}(R)$

In a top-down pass, we now select either C-JMP or C-INS for each ITE tree. This can be done by comparing the cost values in one row of the cost table. The problem is that it is not known in advance *which* row is correct.

At this point, we can exploit the fact, that the *primary* ITE root cannot have a precondition, since there is no surrounding ITE statement. Thus, only the first row in the cost table of the primary root R is relevant. Therefore, we select C-JMP for R, whenever $C_{C-JMP}(R) < C_{C-INS}(R)$, and C-INS otherwise.

Now, there is a kind of domino effect for the subtrees. Let $T' = (R', B_T, B_E)$ be an ITE tree at the next lower level. If we select C-JMP for the primary root R, then according to scheme 1, T' has no precondition. Therefore, it is sufficient to compare $C_{C-JMP}(R')$ and $C_{C-INS}(R')$ to find the best implementation for R', Otherwise, if we select C-INS for R, then T' does have a precondition, and we just need to compare $C^P_{C-JMP}(R')$ and $C^P_{C-INS}(R')$.

Next, let C-INS be selected for R and C-JMP for R' (i.e., scheme 3 is used for R'). Furthermore, let R'' and R''' be ITE tree roots in the then and else blocks of T'. According to scheme 3, R'' has no precondition, and the first row of the cost table for R'' is relevant. In contrast, R''' does have a precondition, so that the second row in its cost table has to be looked up.

In this way, the selection of C-JMP or C-INS for any ITE tree root creates the information which cost table row has to be looked up at the next lower level. Thus, it is possible to optimally decide all implementation schemes in a top-down pass.

Our dynamic programming algorithm is summarized in figs. 6.8 and 6.9. Since each ITE tree root is visited exactly twice, the runtime is linear in the number of ITE statements.

---

algorithm OPTIMIZE_ITE
input: ITE tree T;
begin
   COMPUTECOSTTABLE(T);
   SELECTIMPLEMENTATION(T,false);
end algorithm

algorithm COMPUTECOSTTABLE
input: ITE tree $T = (R, B_T, B_E)$;
begin
   // bottom-up traversal
   for all ITE trees T' in B_T
      COMPUTECOSTTABLE(T');
   end for
   for all ITE trees T' in B_E
      COMPUTECOSTTABLE(T');
   end for
   // let $B_T = (s_1, \ldots, s_n)$ and $B_E = (t_1, \ldots, t_m)$;
   $C(B_T) = \sum_{i=1}^{n} C(s_i)$;
   $C^P(B_T) = \sum_{i=1}^{n} C^P(s_i)$;
   $C(B_E) = \sum_{i=1}^{m} C(t_i)$;
   $C^P(B_E) = \sum_{i=1}^{m} C^P(t_i)$;
   $C_{C-JMP}(R) = S_{C-JMP} + \max(C(B_T) + J, C(B_E) + 2J)$;
   $C_{C-INS}(R) = S_{C-INS} + C^P(B_T \circ B_E)$;
   $C_{C-JMP}^P(R) = S_{C-JMP}^P + \max(C(B_T) + J, C^P(B_E) + 2J)$;
   $C_{C-INS}^P(R) = S_{C-INS}^P + C^P(B_T \circ B_E)$;
   $C(R) = \min(C_{C-JMP}(R), C_{C-INS}(R))$;
   $C^P(R) = \min(C_{C-JMP}^P(R), C_{C-INS}^P(R))$;
end algorithm

Figure 6.8.   Main algorithm for ITE optimization and subroutine for cost table computation

algorithm SELECTIMPLEMENTATION
input: ITE tree $T = (R, B_T, B_E)$;
      bool has_precondition;
begin
  if not has_precondition **then**
    if $C_{C-JMP}(R) < C_{C-INS}(R)$
      then // scheme 1
        select C-JMP for R;
        for all ITE trees T' in B_T
          SELECTIMPLEMENTATION(T',false);
        end for
        for all ITE trees T' in B_E
          SELECTIMPLEMENTATION(T',false);
        end for
      else // scheme 2
        select C-INS for R;
        for all ITE trees T' in B_T
          SELECTIMPLEMENTATION(T',true);
        end for
        for all ITE trees T' in B_E
          SELECTIMPLEMENTATION(T',true);
        end for
    end if
  else // with precondition
    if $C_{C-JMP}^{P}(R) < C_{C-INS}^{P}(R)$
      then // scheme 3
        select C-JMP for R;
        for all ITE trees T' in B_T
          SELECTIMPLEMENTATION(T',false);
        end for
        for all ITE trees T' in B_E
          SELECTIMPLEMENTATION(T',true);
        end for
      else // scheme 4
        select C-INS for R;
        for all ITE trees T' in B_T
          SELECTIMPLEMENTATION(T',true);
        end for
        for all ITE trees T' in B_E
          SELECTIMPLEMENTATION(T',true);
        end for
    end if
  end if
end algorithm

Figure 6.9.   Subroutine for implementation scheme selection

7.   EXPERIMENTAL RESULTS

The experimental evaluation for the above optimization has been performed by generating TI C6201 assembly code for 10 small control-intensive C programs. These have been taken from the ADPCM example in the DSPStone benchmark suite [ZVSM94], as well as from an MPEG C package [MPEG00]. The detailed characteristics of the C sources (number of ITE statements, total size, and nesting level) are mentioned in appendix A (table A.7).

All C codes have been compiled with the TI C6201 C compiler. In addition, the LANCE system described in chapter 8 has been used to convert the C codes into an intermediate representation (IR) in the form of three-address code. A special configuration switch in the LANCE C frontend allows to retain ITE statements in the IR. Then, implementation schemes C-JMP and C-INS have been selected for all ITE statements in the IR by means of the algorithm from fig. 6.8.

From the modified IR, sequential TI C6201 assembly code has been generated. This assembly code has been mapped to parallelized code by means of the TI C6201 assembly optimizer. Finally, the two machine programs generated for each C code have been compared with respect to their worst case execution time and code size.

The machine-dependent parameters J (jump penalty) and K (parallelization factor) have been set to 4 and 3, respectively. The value of J has been chosen smaller than the actual number of branch delay slots in the TI C6201 (5 cycles), in order to accommodate the fact, that sometimes delay slots are filled with useful instructions. This generally results in a lower effective jump penalty.

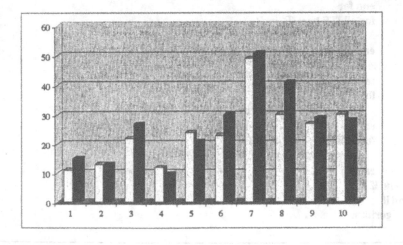

Figure 6.10.   Worst case execution time (instruction cycles) for 10 control-intensive C code examples

The execution time results are summarized in fig. 6.10. The left columns correspond to our optimization algorithm, while the right columns show the results obtained with the TI C compiler. There are three cases (4, 5, and 10), where the TI compiler generates slightly faster code. In these cases, our algorithm has obviously chosen a suboptimal combination of ITE implementation schemes. This is due to the limited accuracy of our estimation technique for the effect of block merging.

In total, however, the code generated by our ITE optimization technique is 7 % faster on the average. The maximum speedup in these experiments has been 27 %.

Like in the case of the instruction scheduling technique from chapter 4, this performance gain is achieved at the expense of an increased code size. On the average, the code generated by our algorithm is 15 % larger than the code generated by the TI C compiler. This is due to the additional instructions needed for computation of conditions when using C-INS in nested ITE statements.

Tables with detailed results on worst case execution time and code size can be found in appendix A (tables A.8 and A.9).

8.   SUMMARY

With the trend towards deeply pipelined and highly parallel VLIW architectures for multimedia processors, usual jump instructions cause significant performance problems, since jumps cause pipeline stalls and limit the freedom for instruction scheduling. As a consequence, recent VLIW processors are equipped with conditional instructions, which allow to generate code for ITE statements without the need for jump instructions. So far, there are very few systematic techniques for utilization of conditional instructions in a compiler, in particular for nested ITE statements. The technique proposed in this chapter operates and the intermediate representation level and uses dynamic programming to obtain an optimized simultaneous selection of either C-JMP or C-INS for ITE statements at all nesting levels. The performance benefits for a real-life processor have been demonstrated experimentally. The results could probably still be improved by using better estimation techniques which also incorporate schedulability information about basic blocks. The core of our technique, however, is the efficient dynamic programming algorithm for ITE trees, which allows to largely separate the actual ITE optimization from machine-specific estimation techniques.

Chapter 7

FUNCTION INLINING UNDER CODE SIZE CONSTRAINTS

In this chapter we consider a performance optimization technique working at the source code level (fig. 7.1). This technique is based on *function inlining*, which is a well-known concept in compiler construction. Function inlining eliminates function call overhead by replacing function calls with copies of function bodies. While this optimization generally results in higher performance, it also increases the code size.

The latter effect can be a problem for embedded systems with a limited amount of program memory. Current techniques for selection of inlined functions are based on simple local heuristics. They result in unpredictable code size and are thus not well-suited for embedded processors. The technique presented in this chapter follows a different approach. Inlined functions are selected in such a way, that the performance is optimized under a global code size limit. We describe a branch-and-bound algorithm that allows to quickly explore the large

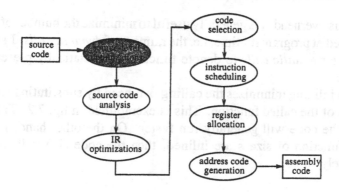

Figure 7.1. Source-level optimizations in the compilation flow

149

search space, and we present an application study for a TI C6201 processor in order to measure the performance increase for a real-life application.

1. FUNCTION INLINING FOR PERFORMANCE OPTIMIZATION

Frequently required subroutines are usually encapsulated into *functions* at the source code level. The extensive use of functions results in better readability of source code and also ensures that machine code for a function appears only once in the compiled program. Furthermore, functions can be put into libraries which allow for software reuse. Therefore, the use of functions in application programming is very popular. On the other hand, an excessive use of functions degrades program performance. This is due to the fact, that functions cause some overhead in the machine code:

Register saving: Live registers, which may be overwritten in a called function must be saved before the function call takes place and must later be restored.

Parameter passing: Function parameters have to be passed to the called function, which requires pushing parameters onto the stack or loading dedicated registers.

Stack frame allocation: A stack frame for the local variables and spill locations of the called function must be allocated and must later be cleaned up. This requires extra instructions for stack pointer manipulation.

Call and return: Special instructions have to be executed for calling and returning from functions, which frequently also require a stack access.

Pipeline stalls: Since call and return instructions obstruct the linear control flow in a program, they may result in a significant number of instruction pipeline stalls (see also chapter 6).

Due to this overhead, it is generally useful to minimize the number of function calls executed at program runtime, i.e. the number of *dynamic calls*. In contrast, we use the term *static calls* to denote function calls statically present in the source code.

Function inlining minimizes the calling overhead by substituting static calls with copies of the called function. This is exemplified in fig. 7.2. The inlined version of the code will generally run faster. On the other hand, if n static calls of a function of size s are inlined, then the code size is increased by approximately[1] $n \cdot s$.

---

[1] The exact code size growth is generally only known after code generation, since numerous detailed effects, such as the amount of register spilling, have to be taken into account. The code size grows only by $(n-1) \cdot s$,

```
/* without inlining */          /* f inlined in g */

int f(int x)                    int g(int x)
{                               {
  return x * 2 + 1;               return (x >> 1) * 2 + 1;
}                               }

int g(int x)
{
  return f(x >> 1);
}
```

Figure 7.2. Example for function inlining

There are two ways of function inlining currently in use in compilers, *semi-automatic* and *automatic*. Semi-automatic means that the compiler user decides which functions should be inlined, while the compiler only performs the "mechanical" part of inlining. This approach can be implemented by introducing a dedicated "inline" keyword as a function qualifier which, for instance, is used in C++ and also in many C compilers. The compiler inlines only the set of functions tagged by the keyword.

In contrast, automatic inlining means that the compiler itself decides which functions should be inlined. For this purpose, a couple of ad hoc rules are commonly used in compilers for general-purpose processors [Much97, Morg98]. These include the following:

- Small functions should be inlined, since their calling overhead might exceed the time needed to execute their body. Some compilers enable the specification of a size threshold value as a compiler option, so as to partially guide the selection of inlined functions by the user.

- Functions with only a single static call should be inlined, since the total code size will generally not be increased.

- Functions called inside loops should be inlined, since this leads to a maximum reduction of dynamic calls.

- Functions that mainly perform a single "case" or "switch" statement on the formal parameters, and which are mostly called with constant actual parameters should be inlined, because it is likely that much of the inlined function

---

if the inlined function is internal (i.e. a "static" function in terms of the C language), because no extra copy of the function code for external calls must be kept.

body can later be eliminated by *constant propagation* and *unreachable code elimination*.

These rules mainly rely on local information and therefore they can be implemented very efficiently in a compiler. However, when using such ad hoc rules it is not guaranteed at all that there is a good trade-off between performance and code size. Like for the other optimizations presented in the previous chapters, for embedded systems it is more favorable to spend more time in code optimization, so as to achieve better solutions.

2.   FUNCTION INLINING UNDER GLOBAL SIZE CONSTRAINTS

Obviously, functions have to be chosen carefully for inlining, in particular if programmers have to cope with limited program memory sizes. While low code size is hardly a major issue for software for general-purpose computers, application programmers for embedded systems usually have to take the size of on-chip program memory into account. Exceeding a certain code size limit means that slow external memory has to be used or that the application cannot be compiled at all because the program address space gets exhausted.

Since we are dealing with embedded processors, our view of function inlining as an optimization problem is that performance should be maximized under a *global* code size limit that captures a complete program. In our case we simultaneously consider all functions contained in a single C file (also called a *translation unit* [KeRi88]).

Consider the situation that a certain program without any inlining has a total size of S. If S is smaller than the code size limit L imposed by the program memory size, then the programmer of an embedded processor is probably interested in the set of functions for which inlining would result in the largest speedup while still meeting the code size limit L. The above local heuristics would simply inline one function after another until the limit L is reached. However, the mutual dependence between inlining of different functions is not taken into account. Therefore, in the context of *constrained* code size, it is unlikely that a good solution will result.

The mutual dependence is a consequence of the fact, that function calls may be nested, and that functions may be called from different places in the source code. The call structure of a program is usually visualized by means of a *call graph* $G = (V, E, B, C)$, where V is a set of nodes, E is a set of directed edges, and B and C are weight functions for V and E.

Each node in V represents one function f_i, and an edge $e = (f_i, f_j) \in E$ denotes that f_i calls f_j. An edge weight $C(f_i, f_j) = n$ means that f_i contains n static calls to f_j. Additionally, we use node weights $B(f_i)$ to denote the *basic size* of a function f_i, i.e., its code size without any inlining of functions called inside f_i.

The total size of a function f_i is given by its basic size plus the sum of the sizes of functions inlined into f_i. These inlined functions may in turn call functions that might be inlined. Thus, the total size of a function eventually depends on the inlining of all functions reachable from f_i in G, whose sizes may also be mutually dependent. Therefore, the decision which functions must be inlined in order to maximize performance should not be made locally. Instead, all functions in a call graph should be taken into account simultaneously. This is the objective of the algorithm presented in the following.

3.   BRANCH-AND-BOUND ALGORITHM

For a given call graph $G = (V, E, B, C)$ and a global code size limit L (measured in bytes), we first identify a set of *inlining candidates* $V' \subset V$. A function $f_i \in V$ is a candidate, if it is not contained in a cycle of G, and if it has at least one predecessor in G. Functions without a predecessor (e.g. the "main" function in C) naturally cannot be inlined anywhere.

If $|V'| = N$, then there are 2^N possible solutions, since each $f_i \in V'$ may or may not be inlined. Due to this exponential search space, it might be necessary to further restrict the set of inlining candidates, so as to achieve acceptable runtimes. However, as will be exemplified later, the algorithm developed in the following is capable of cutting off large portions of the search space without loss of optimality, so that still relatively large candidate sets can be processed.

For the candidate set V' we would like to decide which functions $f_i \in V'$ must be inlined, such that the total code size is not larger than L and the performance is maximal. For sake of simplicity, we assume that if f_i is selected for inlining, then f_i is inlined at *all* positions of static calls to f_i.

This assumption restricts the solution space, but for the remaining solution space we can guarantee that an optimum is found. In our application study this results in an overall better solution than heuristic function inlining for an unrestricted solution space.

Our approach to maximize performance is to minimize the total number of dynamic calls. As explained in the previous section, minimizing the number of dynamic calls tends to increase performance due to a reduction of the calling overhead. However, there is a point of "saturation", where further inlining does not result in higher performance.

Moreover, further inlining beyond this point can even slightly degrade performance due to subtle effects. For instance, in an experimental study [CHT92] it has been found that inlining may increase the number of pipeline stalls. Further sources of performance degradation may be a higher register pressure, caching effects, and less effective instruction scheduling for large basic blocks. Nevertheless, as will be shown later, minimizing the number of dynamic calls can still be used to maximize performance by performing an additional search procedure.

Besides the inlining candidate set, our algorithm needs three sets of input data:

- For any function $f_i \in V$ the *basic size* $B(f_i)$. These values can be determined by compiling the source program once without any inlining and inspecting the assembly listing.

- For any function pair (f_i, f_j) the number $C(f_i, f_j)$ of *static calls* from function f_i to f_j. These values can be determined by a simple source code analysis tool or even manually.

- For any function $f_i \in V$ the number $D(f_i)$ of *dynamic calls*. These values are obtained by profiling for a typical set of input data.

We represent a solution by an *inline vector* $IV = (b_1, \ldots, b_M)$, with $M = |V|$, where

$$b_i = \begin{cases} 0, & \text{if} \quad f_i \in V - V' \\ 0, & \text{if} \quad f_i \in V' \quad \text{and } f_i \text{ is not inlined} \\ 1, & \text{if} \quad f_i \in V' \quad \text{and } f_i \text{ is inlined} \end{cases}$$

Since inlined functions are never called, the total number of dynamic calls for some inline vector IV is given by

$$D(IV) = \sum_{f_i : b_i = 0} D(f_i)$$

Trivially, the theoretical minimum value of D is obtained, if all candidate functions are inlined. However, such a solution is unlikely to meet the code size limit in practical cases.

The total code size of a single function f_i for a given IV is calculated from its basic size, the total size of functions inlined into f_i, and the number of static calls from f_i for such functions:

$$S(f_i) = B(f_i) + \sum_{f_j : b_j = 1} C(f_i, f_j) \cdot S(f_j)$$

The recursion in the definition of $S(f_i)$ terminates at the leaf functions, whose total size is equal to the basic size. The total code size of a complete program for a given IV is

$$S(IV) = \sum_{i=1}^{M} S(f_i)$$

Note that the code size computed in this way is not exact but represents an estimation, since the detailed effects of function inlining on code size are only known after code generation. However, as we will show experimentally, this estimation appears to be sufficiently accurate in practice.

We call an inline vector IV optimal, if $D(IV)$ is minimal and the total code size $S(IV)$ is not larger than the code size limit L. All inline vector bits b_i for which $f_i \in V - V'$ are constantly zero. In the above definitions these are only included for sake of a simpler notation. All bits b_i with $f_i \in V'$ must be set to either 0 or 1.

At the beginning of the optimization, we set all these bits to "x", to denote that their value is yet undecided. We call an inline vector IV containing x-bits a *partial inline vector*. Without loss of generality (since the function indices can be chosen arbitrarily) we assume that any partial inline vector is of the form

$$IV = (b_1, \ldots, b_{i-1}, x, \ldots, x)$$

where all bits $b_1, \ldots, b_{i-1}$ are already set to 0 or 1. Our algorithm for computing the optimal inline vector IV performs a branch-and-bound search, which is based on the following problem analysis.

Given an arbitrary partial inline vector $IV = (b_1, \ldots, b_{i-1}, x, \ldots, x)$, we would like to decide one of its x bits, say b_i. This means, we determine whether or not function f_i must be inlined. We first check, whether f_i could be inlined without violating the code size limit L. When inlining f_i, bit b_i is set to 1, and the total code size is at least

$$S_{lo} = S((b_1, \ldots, b_{i-1}, 1, 0, \ldots, 0))$$

since by setting all bits beyond b_i to zero ensures that no further code size increase is incurred due to inlining of functions $f_{i+1}, \ldots, f_M$. Thus S_{lo} provides a *lower bound* on the total code size when inlining f_i.

If $S_{lo} > L$, then inlining f_i cannot result in a valid solution. This means, b_i must be zero, and any partial inline vector leading to an optimal solution has the form

$$IV = (b_1, \ldots, b_{i-1}, 0, x \ldots, x)$$

Conversely, partial inline vectors of the form

$$IV = (b_1, \ldots, b_{i-1}, 1, x \ldots, x)$$

can be cut off from the search space without loss of optimality.

If inlining of f_i does not necessarily imply a constraint violation, we check whether the decision *not* to inline f_i could still lead to an optimal solution. We use a variable $D_{\min}$ to globally store the minimum number of dynamic calls found so far. At the beginning of the optimization, $D_{\min}$ is initialized by

$$D_{\min} = \sum_{i=1}^{M} D(f_i)$$

which is equal to the number of dynamic function calls without any inlining. During the optimization, D_{min} provides an *upper bound* on the number of dynamic calls in the optimal solution.

If f_i is not inlined, then b_i is set to zero, while the bits $b_1, \ldots, b_{i-1}$ are already fixed. Neglecting the code size, the maximum reduction in the number of dynamic calls in this situation is certainly achieved, if all other functions $f_{i+1}, \ldots, f_M$, are inlined. This is represented by the inline vector

$$IV = (b_1, \ldots, b_{i-1}, 0, 1, \ldots, 1)$$

Computing the number of dynamic calls for this inline vector returns a number $D(IV)$. If $D(IV) > D_{min}$, then not inlining f_i obviously cannot lead to an optimal solution. Therefore, only partial inline vectors of the form

$$IV = (b_1, \ldots, b_{i-1}, 1, x \ldots x)$$

can be optimal, while partial inline vectors with $b_i = 0$. can be cut off from the search space.

Note that it is also guaranteed that a valid solution (i.e. a complete inline vector) for $IV = (b_1, \ldots, b_{i-1}, 1, x \ldots x)$ does exist: The minimum possible code size for IV is obtained, if all bits $b_{i+1}, \ldots b_M$ are set to zero, and we have already verified that

$$S((b_1, \ldots, b_{i-1}, 1, 0, \ldots 0) \leq L$$

In the worst case, neither $b_i = 1$ nor $b_i = 0$ can be excluded in advance. In this case, we recursively compute the complete optimum inline vectors for the partial vectors

$$IV_0 = (b_1, \ldots, b_{i-1}, 0, x, \ldots, x)$$

and

$$IV_1 = (b_1, \ldots, b_{i-1}, 1, x, \ldots, x)$$

The best of these two vectors can be easily determined by comparing $D(IV_0)$ and $D(IV_1)$.

A pseudo code notation of our branch-and-bound algorithm is shown in fig. 7.3. It reads a partial inline vector and returns the optimal complete inline vector. Before calling OPTINLINEVECTOR, we initialize D_{min}, and we construct a partial inline vector

$$IV = (b_1, \ldots, b_{M-N}, x, \ldots x)$$

where all bits for the $M - N$ non-candidate functions are constant zeros, while all bits for candidate function in V' are set to x.

```
algorithm OPTINLINEVECTOR
input: partial inline vector IV = (b₁,...,bₘ) ∈ {0,1,x}ᴹ;
output: (complete) inline vector ∈ {0,1}ᴹ;
begin
  if IV contains no "x" bits then // IV is a complete inline vector
    if S(IV) ≤ L and D(IV) < Dmin then Dmin := D(IV);
    end if
    return IV;
  end if
  i := first index in IV for which bᵢ = x; // IV is a partial inline vector
  IV₁ = (b₁,...,bᵢ₋₁,1,0,...,0);
  if S(IV₁) > L then // fᵢ must not be inlined
    IV' = (b₁,...,bᵢ₋₁,0,x,...,x);
    return OPTINLINEVECTOR(IV');
  else
    IV₀ = (b₁,...,bᵢ₋₁,0,1,...,1);
    if D(IV₀) > Dmin then // fᵢ must be inlined
      IV' = (b₁,...,bᵢ₋₁,1,x,...,x);
      return OPTINLINEVECTOR(IV');
    else // check both possibilities
      IV' = (b₁,...,bᵢ₋₁,0,x,...,x);
      IV₀ = OPTINLINEVECTOR(IV');
      D₀ = D(IV₀);
      IV' = (b₁,...,bᵢ₋₁,1,x,...,x);
      IV₁ = OPTINLINEVECTOR(IV');
      D₁ = D(IV₁);
      if D₁ < D₀ then return IV₁; else return IV₀; end if
    end if
  end if
end algorithm
```

Figure 7.3.   Branch-and-bound algorithm for computing the optimal inline vector

When calling OPTINLINEVECTOR for the first time with the initial vector
IV, the first part of the algorithm is skipped, since IV is a partial, not a complete,
inline vector. This is important, since the test whether

$$S(IV_1) > L$$

guarantees that any further recursive calls of OPTINLINEVECTOR can only
return valid solutions. In an extreme case, this is achieved by setting all bits
$b_{M-N+1}, \ldots, b_M$ to zero, so that no functions are inlined at all.

The remainder of the algorithm implements our above discussion to construct the optimal inline vector, while cutting off parts of the search space at an early point of time whenever possible. At the leaves of the recursion tree, complete inline vectors are passed to OPTINLINEVECTOR. For such a vector IV, the number $D(IV)$ of dynamic calls is computed, and the upper bound D_{min} is updated, if IV is valid and represents a new optimum. Finally, IV is returned without a change.

The worst case runtime complexity of our algorithm is exponential in N, i.e., the size of the candidate function set. However, the two ways of search space pruning we have introduced ensure that the actual runtime typically is much lower. In the following, we use an application study to quantify this, and to show how algorithm OPTINLINEVECTOR can be used as a subroutine to maximize performance rather than just minimizing the number of dynamic calls.

4.   APPLICATION STUDY

In order to evaluate the proposed technique, we have performed an application study for a complex DSP application from the area of mobile telephony: a GSM speech and channel encoder. This application is specified by 7,140 lines of C code, comprising 126 different functions. Out of these, 26 functions are dedicated "basic" functions for certain arithmetic operations. These basic functions, which partially call each other in a non-cyclic scheme, are relatively small and are frequently called from other functions. Therefore, the basic functions were the natural candidates for function inlining in this case. As a target processor, we have used the TI C6201 VLIW DSP together with TI's corresponding ANSI C compiler and instruction set simulator. For such a processor, function inlining may be expected to be particularly important due to the high performance penalty in case of control hazards in the instruction pipeline.

We have used the TI compiler to determine the basic code size $B(f_i)$ of all functions f_i by compiling the source code once without any inlining. A simple source code analysis tool has been used to determine the number of static calls $C(f_i, f_j)$ for all function pairs (f_i, f_j). Finally, the number $D(f_i)$ of dynamic calls for each function f_i have been determined by profiling for a given input speech data file.

Without any inlining, the TI compiler generated code with a size of 67,820 bytes. The initial number of execution cycles as determined by simulation has been 27,400,547.

Considering this number of cycles as 100 %, we have arbitrarily allowed for an increase in code size up to 150 %. The margin between the initial size and the new limit has been exploited for inlining, so as to minimize the number of dynamic function calls. We have already explained that minimizing the number of dynamic calls does not necessarily also minimize the number of execution cycles. Therefore, in order to actually maximize performance, we have performed an interval search for code size limits between 100 % and 150 % in steps of 5 %.

For each concrete code size limit L, algorithm OPTINLINEVECTOR has been applied to identify the optimum set of inlined functions among the 26 candidate functions. Then, we have exploited the semi-automatic inlining capability of the TI C6201 C compiler to tag the respective sets of selected functions with an "inline" keyword, which enforces their inlining in the compiled code. Finally, the resulting code size has been determined, and the number of execution cycles for the compiled program have been measured by simulation. Fig. 7.4 summarizes this procedure.

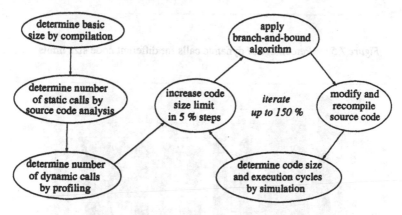

Figure 7.4. Methodology of application study

As can be expected, our first observation was that the number of dynamic calls monotonically decreases with the increase in code size (fig. 7.5), since more program space permits a higher degree of function inlining. For $L = 150$ %, the number of dynamic calls has been reduced to 17 % as compared to the initial solution without any inlining.

However, by simulation we found that the execution cycles do not monotonically decrease, as shown in fig. 7.6. Only up to $L = 110$ % a larger code size directly results in higher performance. Beyond 110 %, the execution cycle numbers vary around 70 % as compared to the initial solution, which indicates the saturation effect mentioned above. The absolute minimum (67 %) has been reached for $L = 125$ %.

This result is comparatively good, since the TI compiler when using its automatic heuristic inlining techniques achieves an execution cycle reduction to only 97 %. In this case, it generates a code size of 106 % compared to the solution without inlining. Obviously, our approach is more effective, since even for a comparable code size limit of 105 %, the execution cycles are reduced to 88 %.

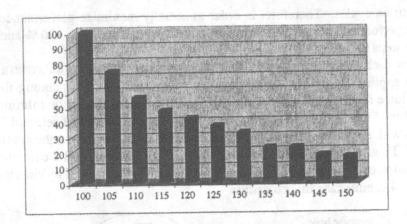

Figure 7.5.   Percentage of dynamic calls for different code size limits

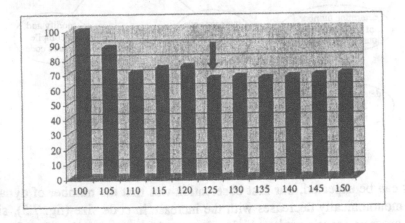

Figure 7.6.   Percentage of execution cycles for different code size limits (arrow shows the absolute minimum)

Our next investigation concerns the code size itself. As we use an estimation $S(IV)$ of the total code size during optimization, the accuracy of the estimation must be analyzed, and it also has to be ensured that the real code size does not exceed the limit. The corresponding experimental results are shown in fig. 7.7. As can be seen, the estimated code size is very close to the real code size determined after code generation, with a maximum deviation of 3 %. Moreover, the code size limit is never exceeded, since the real size is slightly smaller than the estimated size in all cases.

Finally, we consider the efficacy of search space pruning in algorithm OPTIN-LINEVECTOR. Since there are 26 basic functions, the total search space has a size of 2^{26}. Fig. 7.8 shows the percentage of the number of recursive calls to

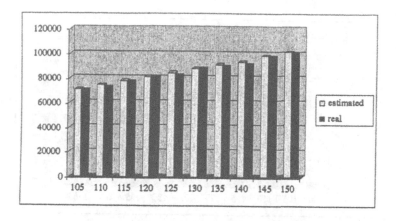

Figure 7.7. Estimated versus real code size

OPTINLINEVECTOR, which indicates the fraction of the search space actually explored. The efficacy critically depends on the tightness of the code size limit. For $L = 105$ %, only 2 % of the search space are visited, because the relatively low limit allows to cut off many partial inline vectors very early. On the other hand, a loose limit reduces the opportunities for search space pruning. For $L = 150$ %, the explored search space amounts to 31 % of the total space.

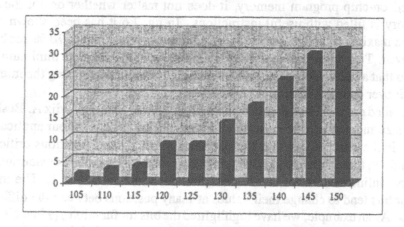

Figure 7.8. Percentage of calls to OPTINLINEVECTOR as compared to 2^{26}

This effect can also be observed when considering the absolute runtime of OPTINLINEVECTOR (fig. 7.9). For $L = 105$ % the algorithm terminates after approximately one minute of CPU time, while for $L = 150$ % about 20 minutes are required.

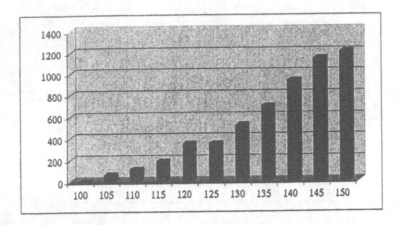

Figure 7.9.   Runtime requirements (CPU seconds) of algorithm OPTINLINEVECTOR

However, the actual bottleneck in our application study was the TI 6201 instruction set simulator, which took between one and two hours per simulation run. In order to avoid very large runtimes due to repeated simulation, our optimization algorithm can also be applied as a stand-alone routine while only considering the maximum available code size. This is reasonable, since for a fixed on-chip program memory, it does not matter whether or not the full memory is filled with useful instructions. In fig. 7.6 it has been shown, that for the maximum limit $L = 150$ % a reduction in cycle count to 70 % has been achieved. This number does not deviate much from the absolute minimum (67 %), so that a satisfactory result is also obtained without performing the interval search over code size limits.

Detailed experimental data can be found in table A.10 in appendix A. Besides code size and performance results, an interesting by-product of our application study is the observation, that the optimum set of inlined functions critically depends on the concrete code size limit. This can be seen when considering the optimal inline vectors for the 26 candidate functions in table 7.1. The inline vector bits tend to change their values at many positions between the different limits. As an example, we have highlighted the bits for function f_{17} in table 7.1. For the code size interval [105 %, 115 %], function f_{17} is inlined. However, this changes within the interval [120 %, 130 %], for which the available space is obviously better used for inlining other functions. Nevertheless, there is another interval, [135 %, 140 %], where inlining f_{17} is favorable again.

In total this indicates that among the set of candidate functions there are few functions for which inlining pays off independent of the code size. This motivates the use of our time-intensive, but effective, global approach to function inlining. In contrast, finding the optimum set of inlined functions for a given code size limit manually or based on local heuristics seems to be very difficult.

Table 7.1.   Optimal inline vectors computed by the branch-and-bound algorithm

size limit (%)	inline vector (functions 1-26)
100	0000000000000000	0	000000000
105	0010000000110000	1	110111111
110	1011100101110000	1	111111111
115	1011000000000100	1	000111001
120	1011010010100010	0	110111101
125	1011000000101000	0	100111101
130	0011000000001010	0	100111000
135	1011001000111010	1	110111101
140	1011101111111010	1	111111111
145	1011011010101010	0	110111101
150	1011011000001011	0	110111101

5.   SUMMARY

The source-level optimization technique presented in this chapter can be used for performance optimization under a global code size constraint. This meets the demands of embedded systems, where there is often a small and fixed amount of available program memory. Due to their locality, current heuristic inlining techniques do not meet such demands and make no sophisticated trade-off between code size and performance. The proposed algorithm is capable of exactly minimizing the number of dynamic function calls during the execution of a program by selecting a subset of functions to be inlined for a given code size limit. Approximately, this also minimizes the number of execution cycles for the compiled code. In an application study, we have shown how the actual performance can be maximized by embedding our optimization algorithm into an interval search over different code size limits, which accommodates the saturation effect often encountered in function inlining. Even though there is an exponential search space, the presented algorithm can be used for realistic problem sizes, since it typically needs to explore only a fraction of the search space. For the GSM application and the TI C6201 processor, the net result was a maximum performance increase by 33 % by means of optimized function inlining, while giving the user perfect control over the resulting code size.

Chapter 8

FRONTEND ISSUES – THE LANCE SYSTEM

The code optimization techniques presented in the previous chapters mainly refer to what is commonly called the *backend* of a compiler, i.e., the machine dependent part responsible for assembly code generation. However, a significant portion of the total design effort for writing a compiler also goes into the machine-independent parts, i.e., the *frontend*. This includes the source language parser, as well as machine independent optimizations working on the intermediate representation (IR). In most cases, a frontend is required for experimentally evaluating new code generation techniques. Also the transfer of research results into industrial practice can hardly be accomplished without a frontend.

Since the frontend design and the IR being used have some impact on backend techniques, in this chapter we briefly describe how a C frontend called LANCE has been designed in the context of the research described in this book. LANCE provides a common platform both for research on code generation techniques and the construction of complete compilers. It comprises an ANSI C frontend, a collection of IR optimizations, as well as a procedural interface to the IR (fig. 8.1). The system is designed in such a way, that machine-specific backends can be easily integrated.

1. DEMANDS AND RELATED SYSTEMS

The theory of frontends and IR optimizations is well-known and can be found in numerous textbooks, e.g. [AhUl72, ASU86, WiMa95, Much97, Appe98, Morg98]. Therefore, we will focus on software design and implementation aspects. Essentially, our objective is to show how a development platform for C compilers can be built with relatively low manpower. The design decisions for LANCE have been influenced by the following factors:

165

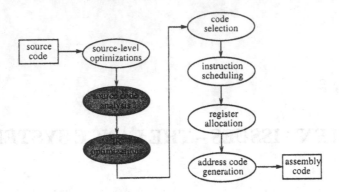

Figure 8.1.   Compilation tasks covered by LANCE

Flexibility.   The required degree of flexibility has been a main motivation for the development of LANCE. In order to be applicable for a variety of target processors, the system must be machine-independent, i.e., it must rely on only very few machine-specific information. In particular, no assumption must be made about the target instruction set and register architecture.

Also the source language itself must be flexible to a certain extent. In some cases, an "extended subset" of the C language is required, which restricts permissible data types, but includes new keywords. This can usually not be achieved when using third-party tools.

In LANCE, all IR optimizations are implemented as "plug-and-play" components that work on a common IR. Due to the mutual dependence of IR optimizations (i.e., some optimizations generate or restrict opportunities for others), the order of executing IR optimizations should be flexible as well. If the highest optimization level is required, then all IR optimizations should be iterated by a script until a fixpoint is reached. This is normally not available in other compilers, which use a fixed order of IR optimizations to limit the compilation time.

Reuse.   It must be possible to reuse the system when writing backends for new target processors. This is achieved by the machine-independence of the IR. In addition, the LANCE system comprises a C++ library with important basic data structures (such as stacks and graphs), functions for IR import and export from/to files, as well as control and data flow analysis functions required in many tools.

Cost.   The development effort should be limited, since the LANCE system primarily serves as a research and development platform within an academic environment. Therefore, efficiency of the compilation process itself has not been a major goal for LANCE, and massive use has been made of generator technology.

Reliability.   Since a compiler frontend together with IR optimizations is a relatively complex software, the implementation strictly follows the principle of modularity. Therefore, C++ has been selected as the implementation language, and all tools within LANCE are separate and use a common set of library functions. The generation of IR code within the C frontend is syntax-directed. In order to facilitate validation, in LANCE version V2.0 an IR has been introduced that is a subset of the C language. In this way, generation and modification of the IR can be validated by compilation with existing compilers.

We found that existing publicly available compiler frontends do not meet all of these demands. Commercial frontends, such as Cosy [ACE00] (which also comprises a backend generator), Cocktail [CoCo00], or EDG [EDG00] offer a high degree of reliability, but they are expensive and not very flexible.

The SUIF system from Stanford University [SUIF00] offers a high degree of flexibility concerning the IR, but it uses a very complex internal representation, motivated by the need to serve different source language frontends. Currently, a completely revised software version, SUIF 2, is under development, which so far is only available as a beta test version. As the SUIF developers have switched to a new C frontend, source language flexibility is presumably limited in practice, and frontend sources have to be commercially licensed.

The Trimaran system [Trim00] is another platform for compiler research, that comes with a C frontend and a collection of IR optimizations. However, it focuses on optimizations for instruction-level parallelism, and therefore it supports only a specific class of parameterizable VLIW processors.

Also existing retargetable compilers turned out to be hardly applicable in the context of code generation for embedded processors. For instance, the lcc compiler [FrHa95] directly generates assembly code from C code. However, it does not support integration of custom code generation techniques and it performs only few IR optimizations. Similar restrictions hold for the commercial retargetable compiler from Archelon [Arch00].

Also the GNU C compiler gcc [GNU00], which has been successfully ported to a number of RISC and CISC targets, is less suitable for embedded processors. This is due to its rigid retargeting mechanism and its limited range of target architectures. For instance, gcc based compilers for DSPs have not been very successful [ZVSM94]. In fact, comfortable retargeting and a wide range of target architectures have never been a design goal in gcc, and a number of users have reported severe problems when attempting to port gcc to an application-specific architecture. A detailed report on experiences with porting gcc to an embedded processor is given in [GuLu95]. Main problems included the

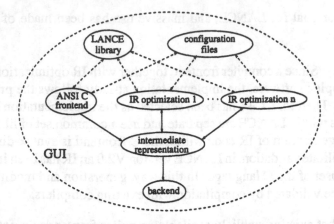

Figure 8.2.   LANCE V1.0 software architecture

macro-based retargeting mechanism, gcc's preference for using registers, and the restriction towards byte-addressable memories.

During the design of backends for different embedded processors, we found that the target architectures encountered in the area of embedded systems show such a large variety, that a retargetable C compiler that directly generates assembly code for arbitrary processor architectures seems impossible, or at least extremely inefficient. Most parts of the compiler backend have to be machine-specific, and only for some subtasks existing generators (such as OLIVE) can be directly reused. Therefore, in contrast to many of the above systems, the LANCE system presented here is strictly machine-independent. This means that backends for new processors have to be designed almost from scratch, but the effort for porting the frontend is minimal. This made it possible to reuse the system for a number of different compiler projects. In the following, we describe some implementation aspects of LANCE.

2.   SOFTWARE ARCHITECTURE

Fig. 8.2 shows the general architecture of the LANCE V1.0 system. The ANSI C frontend generates the IR for a given C source file. A collection of IR optimization tools may read and modify the IR, and write it back into a file. Currently, tools for nine IR optimizations are implemented.

The implementation of these tools (except for the use of conditional instructions, see chapter 6) directly follows techniques from compiler textbooks and thus needs not to be further detailed here. Since the IR format is identical for all tools, new optimizations may be integrated at any time, and target-specific optimization scripts can be specified. After IR optimization, a custom backend may be called to translate the IR into assembly code.

Machine-specific parameters are passed to the tools via configuration files. For the C frontend, for instance, this concerns the bit widths and memory alignment of the different C data types, while for common subexpression elimination the processor-specific interval of constants, which must not be considered as common subexpressions, can be specified.

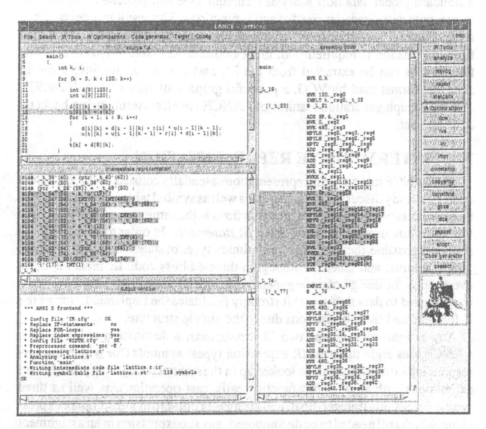

Figure 8.3. Screenshot of the LANCE V1.0 graphical user interface. Here, the assembly code window (right) contains symbolic TI C6201 assembly code.

The C frontend, the IR optimizations, as well as backends make use of a common library of basic data structures, IR I/O and manipulation functions, and flow analysis functions. All tools can be used as shell commands or, more conveniently, can be operated via the GUI (fig. 8.3). The GUI is implemented with the script language Tcl/Tk [Oust94] and has text windows for C code, IR code, and assembly code, as well as a control output window. In order to make the compilation process transparent when developing IR optimizations or backends, the code windows are linked in such a way, that a mouse-click on a line in one window automatically highlights the corresponding lines in the other two windows. Also symbol table information for identifiers and data flow information can be directly displayed.

A special tool in LANCE allows for the display of flow graphs at the IR level. The *control flow graph* visualizes the basic block structure of functions, while the *data flow graph* represents the dependencies between IR statements. We use a standard global data flow analysis technique (see e.g. [Much97] for details) to compute data dependencies for entire functions. The result is a dependence representation in the form of *def-use chains* and *use-def chains*. This data flow information is required both in IR optimizations and in backends. The flow graphs can be extracted from the IR, and can be written into a file in a special format read by VCG, a powerful graph visualization tool [Sand95]. The flow graph generation features of LANCE are also useful during backend development.

3.   INTERMEDIATE REPRESENTATION

The LANCE intermediate representation essentially consists of three-address code, as already discussed in section 1., as well as symbol table information. The three-address code is structured into functions in the same way as in the C source file. Each function consists of a list of *IR statements*. In order to keep the IR as simple as possible, there are only four primary types of statements: assignments, jumps, returns, and labels. Although a three-address code IR incurs a slight overhead in the design of code generators (since IR statements later have to be re-combined to data flow trees) it strongly facilitates the implementation of the C frontend and IR optimizations due to the simple structure.

Any assignment contains two *IR expressions*, a destination and a source. LANCE uses eight different IR expression types: symbols (for which the corresponding information can be looked up in the symbol table), unary and binary expressions with an operator, function calls, cast operations, as well as three types of constants (integer, float, and string). In order to avoid complex expressions, which still needed to be decomposed, any subexpression in an assignment source must be a symbol, a constant, or a pointer dereference. The latter is considered as a special unary expression. Likewise, assignment destinations must be symbols or pointer dereferences. The latter translate to STORE instructions in the assembly code.

Normally, any high-level control structures (if-then-else, for, while, do, switch, break, continue) are replaced by corresponding low-level constructs using conditional and unconditional jump statements (see chapter 6). However, in order to enable certain optimizations, the frontend can be configured in such a way, that high-level statement types are retained in the IR. These must then be lowered in a later phase, so as to make the IR optimization tools and flow analysis library functions applicable. Also all implicit address computations in C, such as for array and structure accesses, as well as code for initialization of local variables are explicitly inserted into the IR in order to simplify the design of backends.

For the most recent version of the LANCE system (V2.0), the IR has been slightly redefined to be a *subset of ANSI C*, which we call IR-C. The structure of IR-C is the same as described above. However, there are several advantages of using IR-C as compared to a custom IR format. The most important one is that validation of the C frontend and IR optimizations is directly supported, since the IR is *executable*: Given a C source program P_1 and a corresponding IR-C program P_2 generated by the C frontend, one can use an existing native compiler on the development host (e.g. gcc on a workstation) to check the behavioral equivalence of P_1 and P_2 for a set of test input data.

The use of a C subset as an IR sometimes requires special techniques. For instance, consider an array access x = A[i] to an integer array A on a machine with 32-bit integers and a byte-addressable memory. The bit widths of C types and addressable memory units are passed as configuration parameters to LANCE. In order to simplify backend design, all address arithmetic and cast operations are made explicit already by the C frontend. By definition, A[i] refers to a memory address computed by adding the array base address A and the array index i scaled by $32/8 = 4$. Thus, at first glance, we could replace this by the low-level C code

```
int* t1, *t3;    // auxiliary variables
int t2;

t1 = A;          // take array base address
t2 = i * 4;      // scale index
t3 = t1 + t2;    // compute address of A[i]
x = *t3;         // memory access
```

However, this code is not equivalent to the original one. The reason is, that in statement t3 = t1 + t2 still an implicit scaling of t2 by 4 is performed, since t1 is an integer pointer. Therefore, when compiling this code with a C compiler, the memory access via *t3 will not take place to the desired location.

In LANCE, we solve this problem by temporarily performing all address arithmetic only on character pointers. By definition, the C type char has a size of one byte and therefore should always fit into a single memory word. Thus, no implicit scaling takes place when adding integers to character pointers, and the following IR-C code shows the correct behavior:

```
char* t1, *t3;    // auxiliary variables
int t2;
int* t4;

t1 = (char*)A;    // cast base of A to a char pointer
```

```
t2 = i * 4;          // scale index
t3 = t1 + t2;        // compute address of A[i]
t4 = (int*)t3;       // cast back to an integer pointer
x = *t4;             // memory access
```

Additional cast operations in the IR do not cause overhead in generated machine code, since a pointer-to-pointer cast requires no extra instructions at the assembly level.

Due to the replacement of all implicit address computations, a generated IR-C program is equivalent to an original C program only for a specific machine. For instance, an IR-C program generated for a Unix workstation generally will not run on a Linux PC, due to different type bit width and alignment parameters. However, determining the required parameter configuration for a certain machine is rather straightforward. In case a native compiler is already available (which is typically the case for a compiler development host), the parameters can even be automatically determined by compiling and running special "configuration detection" programs. When developing a new cross-compiler, the machine parameters, within certain limits, are anyway chosen by the compiler developer himself.

Besides the support for validation, further benefits of the C-based IR in LANCE include the following:

- LANCE can be employed for certain source-to-source transformations (in our case IR-C to IR-C) which, for instance, can be applied to address code optimization [GGMC00]. In fact, the IR optimizations in LANCE can be viewed as low-level C-to-C transformations.

- From a software engineering viewpoint, it is important that the same frontend can be used for parsing C source files and IR files, although for the latter only a "light" version without error checking and decomposition of complex expressions is required.

- The use of IR-C also minimizes the effort for new LANCE tool developers to get used to the IR structure, since programmers familiar with the C language can understand the IR immediately.

The disadvantage of IR-C is a lower efficiency of the compilation process, since IR-C definitely is not the most compact IR format. The concrete definition of IR-C is rather straightforward and is therefore omitted here. Appendix B provides a complete example in order to outline the correspondence between C and IR-C.

4.   ANSI C FRONTEND

The design of a C frontend is a relatively complex task, even though tools like lex and yacc [MaBr91] facilitate the generation of scanners and parsers for lexical and syntax analysis, respectively. Yacc-compliant C grammar specifications are even publicly available [Dege95]. The main problem lies in the semantical analysis, which deals with the meaning of a source program rather than its outer form. As mentioned in [FrHa95], *the compiler writer must understand even the darkest corners of the C language*, even though many language features are hardly used in practice. For instance, it may be somewhat surprising that the C assignment

```
A[1] = f(x);
```

can be equivalently written as

```
1[A] = (**********************f)(x);
```

due to the type conversion rules in C and commutativity of the array access operator [].

4.1   ATTRIBUTE GRAMMARS

A convenient way to cope with the large and partially complicated set of semantic rules in programming languages such as C is the use of *attribute grammars*. These allow for a syntax-directed, and thus very clean and modular, analysis. We will just give a brief description from a practical viewpoint here, and we assume that the reader is familiar with the basics of context-free language parsing. Our purpose is to outline how an attribute grammar for ANSI C can be specified, including the generation of the C-based IR. Theoretical background information on attribute grammars can be found in [ASU86, WiMa95].

The syntax of a context-free language is normally captured by a context-free grammar $G = (\Sigma_N, \Sigma_T, R, S)$, where Σ_N is a set of nonterminals, Σ_T is a set of terminals, R is a set of rules, and $S \in \Sigma_N$ is the start symbol. Attribute grammars are constructed "on top" of a context-free grammar. In an attribute grammar, all symbols $x \in \Sigma_N \cup \Sigma_T$ are annotated with an attribute set $A(x)$. An attribute $a \in A(x)$ can be considered as a container for semantical information about a symbol. In practice, a is nothing but a variable.

Typical attributes include: names and types of variables, *environments* (nested symbol tables) in which expressions are evaluated, source code information for debugging purposes, and Boolean attributes indicating whether a semantical error has been detected during application of a rule in R.

The value of an attribute $a \in A(x)$ is determined by an *attribute definition* $D(a)$, an equation attached to a grammar rule in which a occurs. Attributes fall into two classes: *synthesized* and *inherited* attributes. An attribute $a \in A(x)$ is synthesized, if x is a nonterminal on the left hand side of the rule, and its definition only depends on attributes of grammar symbols on the right hand side of the same rule. As an example, consider the C grammar rule (given in yacc syntax)

```
unary_expression : SIZEOF '(' type_name ')'
```

where unary_expression and type_name are nonterminals, while SIZEOF is a terminal. Suppose, type_name has an attribute T, denoting its C type, and unary_expression has an attribute V, denoting its numerical value. Since V only depends on the number of memory words occupied by T, V is a synthesized attribute.

Synthesized attributes can be used to pass information "upwards" in the parse tree generated by syntax analysis. Typically, such attributes are used to compute new information within the parse tree that combines information from subtrees, e.g., the type of an expression can be determined based on the types of its subexpressions.

Conversely, inherited attributes allow for passing information "downwards" in the parse tree. This mechanism is very useful, because it allows to process grammar rules within their current context. A good example is the specification of "pointer generation" in C, which reads [KeRi88]: *If the type of an expression [...] is "array of T", for some type T, then the [...] type of the expression is altered to "pointer to T". This conversion does not take place if the expression is an operand of the unary & operator [...].* The C grammar rule for parsing occurrences of variables in expressions is

```
primary_expression : IDENTIFIER
```

The type T of nonterminal primary_expression must be stored in some attribute. Normally, it is equal to the type T' of terminal IDENTIFIER, which can be looked up the the expression environment. If IDENTIFIER happens to represent an array of elements with type T'', then pointer generation has to be applied to make T a pointer to T''.

However, according to the above semantical rule, an exception is that primary_expression is a subexpression of a unary expression containing the address operator "&", but during parsing this situation is not yet visible when processing the above grammar rule.

A solution is to use an additional Boolean attribute, say *do_pointer_generation*, which specifies whether or not pointer generation must be applied to the type of primary_expression as usual, or whether an exception due to the "&" operator occurs. The kind of operation applied to primary_expression is only known when using the C grammar rule

```
unary_expression : unary_operator primary_expression
```

which parses expressions with a unary operator. However, during syntax analysis this rule is applied *later* than the previous one, since the reduction of the identifier to primary_expression already must have taken place. The use of the inherited attribute *do_pointer_generation* allows to break this cyclic dependence. We can define *do_pointer_generation* for primary_expression to be "true", if and only if nonterminal unary_operator represents the "&" operator. Due to the inheritance, the value of *do_pointer_generation* can already be used when determining the type of primary_expression in the rule primary_expression : IDENTIFIER.

Another typical application of inherited attributes is to pass an environment of currently visible object declarations to a C statement list. The body of a C function matches the grammar rule

compound_statement : '{' declaration_list statement_list '}'

In order to check semantical correctness of the statements within the statement list, the set of declared local variables (captured as an attribute of the declaration list), as well as all global variables must be known when parsing the statements. Therefore, a symbol table containing all global variables as well as the local variables must be "inherited" to statement_list.

Also "forward" jumps be means of C goto statements can be conveniently processed with inherited attributes. The problem is, that labels need not to be declared in C, so that the target label of a goto might be parsed only later that the goto statement itself. However, the presence of the target label must be known when checking the semantical correctness of the goto. This can be resolved by using two environment attributes, E_1 and E_2, for a compound statement. In a first pass, E_1 is used to collect all labels within statement_list. Then, in a second pass, E_2 is set to E_1 by means of inheritance, and E_2 serves as the regular environment for statement_list. The point is, that no code needs to be written to implement those two passes, but that the passes are implicit in the attribute definitions.

4.2   THE ATTRIBUTE GRAMMAR COMPILING SYSTEM OX

The well-known UNIX tool yacc allows for automatic generation of parsers from a context-free grammar specification. More precisely, yacc generates C code from a Backus-Naur specification of a grammar, where the generated code comprises a dedicated function "yyparse" for parsing an entire input file. Yacc usually cooperates with lex, which is capable of generating scanners that transform input files into a stream of *tokens*. These tokens must be identical to the terminal set of the context-free grammar used by yacc. From a theoretical viewpoint, lex generates finite state machines, while yacc generates stack machines.

Lex and yacc allow for a convenient implementation of syntax analyzers. Yacc also shows limited support for attribute grammars, since it allows to specify *action code* (similar to OLIVE, chapter 3) in the form of C code for grammar rules. The action code is executed each time its grammar rule is applied during parsing. By means of the action code, each rule may "return" a single attribute, i.e., it can pass some information upwards in the parse tree.

The limitation of this mechanism is that only synthesized attributes can be used. In order to perform semantical analysis, one has to use the synthesized attributes to build up a parse tree data structure, on which type checking, symbol table book-keeping, and IR code generation have to be performed in later traversals. Thus, a clean syntax-directed source code analysis is not possible.

Due to these complications, we have used the OX tool for developing the LANCE ANSI C frontend. OX can be regarded as a generalization of lex and yacc towards attribute grammars [Bisc92]. The functionality of OX and its cooperation with other tools are sketched in fig. 8.4. OX reads attributed scanner and parser specifications and generates regular lex and yacc input files, which can then be processed further as usual with lex, yacc, and a C compiler.

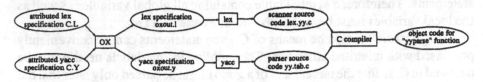

Figure 8.4.   Use of the attribute grammar compiling system OX

The close relation of OX to yacc is a significant practical advantage with respect to the tool chain for compiler development. The syntactic parts of analyzers built with OX are identical to yacc. Only the attribute definitions have to be specified with special constructs, however with a close relation to the C language. In fact, arbitrary C code can be used in attribute definitions. OX places the C code required for evaluating attribute definitions into the yacc action functions, augmented with special code that ensures an evaluation in the right order. The user only has to be aware of the semantical differences between attributes and usual variables. An attribute must have at most one definition per rule, and cyclic dependencies between attributes must be avoided.

A very convenient feature of OX is its capability of specifying *auto-synthesized* and *auto-inherited* attributes. Since the number of attributes may be quite high (as much as 20 attributes per grammar symbol in LANCE), the detailed specification of all attribute definitions for each grammar rule would result in a huge grammar specification. However, many attribute definitions are in fact redundant, in the sense that attributes values are frequently just passed through the parse tree without any modification. For auto-synthesized and auto-inherited attributes, OX is capable of automatically inserting the corresponding attribute definitions, whenever the definition can be uniquely derived from the context.

OX processes arbitrary attribute grammars, with the restriction that synthesized attributes of terminals (or tokens) must not depend on any other attributes. This restriction is normally not a problem, because token attributes mostly consist of lexical information, which can be directly extracted from the input file. The only exception is the required distinction between C identifiers and type names. From the scanner's viewpoint, both are character strings and cannot be distinguished. However, the parser needs to distinguish tokens for identifiers and type names, because otherwise the context-free grammar gets ambiguous. Therefore, an inherited attribute, which signals the context of a character string to the scanner, would normally be required, but is not supported by OX. This problem can be circumvented by using the usual yacc actions in addition to the attribute definition mechanism offered by OX.

4.3   IR GENERATION

An important concept in the C attribute grammar used in LANCE is that also the IR generation itself is exclusively handled in a syntax-directed manner via attributes, so as to increase the modularity of the implementation. All C grammar constructs that may require generation of a piece of IR code (i.e. expressions and statements) have two inherited attributes env_in (the input environment) and IR_in (the previous list of IR statements), as well as two synthesized attributes env_out (the output environment, containing possibly required new auxiliary variables) and IR_out (the updated IR statement list). The latter two are defined by means of a dedicated IR generation function specific to the current grammar rule. In each grammar rule, a piece of IR code may be appended to the previous list of IR statements, and new auxiliary variables may be inserted into the current environment.

As an example, we consider IR generation for a for-loop in C. Such a loop in general has the form

```
iteration : FOR '(' init ';' test ';' update ')' statement
```

where init, test, and update are expressions for loop initialization, testing the loop exit condition, and updating a loop variable, respectively. Nonterminal statement represents the loop body and denotes any type of C statement, in this case mostly a compound statement. The IR code for the four components of the for-loop are supposed to be stored in the synthesized attributes init.IR_out, test.IR_out, update.IR_out, and statement.IR_out. Then, attribute IR_out of nonterminal iteration is synthesized in such a way such that it contains IR code of the following form:

```
          <init.IR_out>                  // IR code for init
L_start:                                 // loop start label
          <test.IR_out>                  // IR code for condition testing
          if (test.cond) goto L_end;     // conditional jump to end label
          <statement.IR_out>             // IR code for loop body
L_next:                                  // loop continuation label
          <update.IR_out>                // IR code for variable updates
```

```
          goto L_start          // backward jump to loop start
L_end:                          // loop end label
```

Here, attribute test.cond is supposed to store the name of a Boolean auxiliary variable containing the result of the exit condition test. Essentially, the IR generation function for the above example rule merely needs to concatenate the IR code fragments of the for-loop components in the right order, while inserting labels at the correct positions. However, there is also a top-down flow of information by means of inherited attributes, because when parsing the loop body represented by statement, the concrete names of labels L_next and L_end have already to be known. The reason is that within the loop body, there may be continue or break statements, in which case a jump to those labels has to be performed. Likewise, statement needs to have an inherited Boolean attribute is_loop_body, so that during semantical analysis of the loop body it can already be decided whether or not the statement context allows for continue or break statements.

Besides IR code, i.e. statement sequences, also symbol table information is kept in the IR. The C language allows for nested local scopes for variable and type names in functions. During semantical analysis, the LANCE C frontend uses nested hash tables to maintain environments for C statements and expressions. When emitting the internal IR format into a file in IR-C syntax, the local symbols tables of functions are flattened, so that only two environment levels remain: a global environment and one local environment per function. In this way, compound statements with local variables can be avoided in the IR, which simplifies its structure. Name conflicts are avoided by giving all local symbols are unique numerical suffix. Naturally, global names are not changed, since otherwise linking with other C modules would be impossible. In addition, all type definitions are resolved, and all structure declarations are assigned a unique tag name and are made global.

5.   BACKEND INTERFACE

After generation and optimization, IR files are normally passed to a processor-specific backend. In order to bridge the gap between the three-address form of the IR and the data flow trees (DFTs) usually required by code generator generators such as OLIVE (see chapter 3), the LANCE system also comprises a backend interface. This interface transforms a three-address code IR into a behaviorally equivalent sequence of DFTs. The terminal set of the DFTs is fixed and machine-independent. Therefore, the backend interface can be reused for all different target processors, and the generated data structures are fully compliant with the formats required by OLIVE or IBURG.

The basic technique in DFT generation is *substitution*: Under certain conditions, uses of variables can be substituted by their corresponding definitions, which is a simple application of data flow analysis. In this way, expressions are step-by-step enlarged, so as to form "complete" DFTs. However, care must be taken during determination of DFT boundaries, in order to prevent undesired side effects. Essentially, this amounts to the correct identification of DFT root statements. In the LANCE backend interface, an assignment is considered as a DFT root, if

1. it writes a value to memory, i.e., it is a store operation, a write to a function argument, or a write to a variable to be kept in memory, or

2. it writes to a variable that has multiple occurrences on the left hand side of assignments, or

3. it has a function call on its right hand side, or

4. its left hand side has multiple uses (common subexpression), or

5. its left hand side has at least one use in a different basic block (in which case it has to be passed via a register), or

6. is has a loop-carried use.

Besides its reusability, the main advantage of the machine-independent backend interface in LANCE is that the same validation technique as for the IR optimizations can be applied. The LANCE system comprises a tool for exporting the generated DFTs in C syntax. The generated C files can be compiled and checked in the same way as regular IR-C files, as described earlier.

6.   APPLICATIONS

The LANCE tool concept has proven its applicability in a number of academic and industrial compiler projects. Within the context of the research described in this book, LANCE has been applied for experimental evaluation of several techniques. An example is the use of SIMD instructions (chapter 5). For this purpose, the data flow graph display features of LANCE turned out to be very useful, since the modification of the OLIVE tool for generation of alternative covers needed to be validated. Fig. 8.5 gives an example of a data flow graph with alternative matching rules annotated in the graph nodes.

Further applications of LANCE include prototype compilers for TI 'C5x DSP, M3 DSP [FWD+98], AMS Gepard DSP [AMS00, Scho99], and the ARM7 RISC core. In addition, LANCE is being used in commercial compiler development projects carried out at the ICD technology transfer company [ICD00]. Current ICD projects include the implementation of C compilers for a bit-stream coprocessor designed by Infineon Technologies, and a scalable DSP for mobile applications designed by Systemonic AG [Syst00].

The C frontend of the LANCE V2.0 system is freely available via the World Wide Web [LANC00] for Sun/Solaris, PC/Linux, and PC/Win95 platforms.

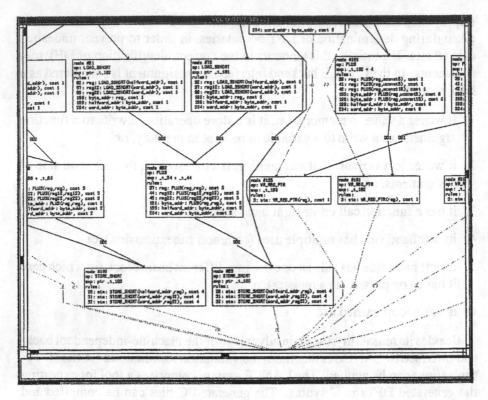

Figure 8.5.   DFG fragment with alternative covers for code selection with SIMD instructions and different inter-node dependencies (VCG display)

The complete system, including IR optimizations and library, is available based on research license agreements or commercially within the context of industrial compiler projects through the ICD.

Chapter 9

CONCLUSIONS

MOTIVATION

The need for higher productivity in embedded system design demands for taking the step from assembly to C level programming of embedded processors. While compiler technology is relatively mature for general-purpose processors, this generally is not the case for embedded processors. The reason is that embedded systems require higher code efficiency than general-purpose systems. Simultaneously, recent embedded processors show specialized architectural features that cannot be well exploited with standard compiler techniques. However, C compilers for embedded processors will only gain more acceptance in industry, if their code quality overhead is low as compared to assembly code. The key concept to overcome these problems is the development of new compiler techniques that take the special constraints in embedded system design into account.

CONTRIBUTIONS OF THIS BOOK

In this book, we have reviewed the state-of-the-art in code optimization for embedded processors, and we have presented a number of new techniques for different processor families. Experimental results for real-life applications and embedded processors indicate the large optimization potential of such techniques. Emphasis has been put on recent VLIW-like multimedia processors. Amongst others, we have presented a first approach to exploitation of SIMD instructions without the need for C language extensions. We expect that VLIW processors in the future will become even more important for embedded systems due to the high performance requirements. In order to put the machine-level optimizations into a wider context, we have also covered source-level tech-

181

niques, such as function inlining, as well as compiler frontend technology as exemplified by the LANCE system.

FUTURE PERSPECTIVES

The optimization techniques presented here mainly refer to performance and/or code size. However, as already mentioned in the introduction, there are further compiler design goals in the area of embedded systems. As an additional optimization goal (possibly contrary to performance and code size), low power consumption is getting more and more important. Existing code optimization techniques should be investigated for their impact on power consumption of software, but the largest benefit may be expected from completely new techniques yet to be developed. From a system-level design perspective, retargetable compilation might be expected to receive renewed interest very soon, due to the need for architecture exploration tools. Retargetable compilers may provide an early feedback on alternative target architectures, while the highest optimization effort is only required once the architecture has been fixed. Moreover, the growing number of application-specific customizable processors (ASIPs) used in system design indicates the need for retargetability. Not only does this hold for compilers, but also for additional software development tools like simulators, assemblers, and linkers.

Appendix A
Experimental Result Tables

Table A.1. Performance of the genetic algorithm (GA) for $(k, 0, 1)$-OA (chapter 2) as compared to the heuristic from [LeMa96b] for different problem parameters (number of address registers k, access sequence length $|S|$, number of variables $|V|$. The numbers refer to the average offset assignment costs of the heuristic (column 4) and the GA (column 5). Column 6 mentions the percentage of costs achieved by the GA as compared to the heuristic (set to 100 %).

k	$|S|$	$|V|$	GOA [LeMa96b]	GA	%
2	50	10	14.3	11.9	83
2	50	20	18.0	16.3	91
2	50	40	10.4	9.8	94
2	100	10	33.0	29.3	89
2	100	50	41.6	41.0	98
2	100	90	11.2	11.0	98
4	50	10	5.7	4.3	75
4	50	20	10.7	7.0	65
4	50	40	9.5	7.1	75
4	100	10	9.8	6.6	67
4	100	50	29.0	27.5	95
4	100	90	9.5	8.2	86
8	50	10	5.0	4.2	84
8	50	20	9.0	5.7	63
8	50	40	11.9	7.1	60
8	100	10	5.0	5.0	100
8	100	50	20.6	15.9	77
8	100	90	12.7	9.0	71
				average	82

183

Table A.2.   Performance of the GA for $(k, m, 1)$-OA (chapter 2) as compared to the heuristic from [LeMa96b], combined with post-pass exploitation of modify registers. The structure of the table is the same as in table A.1.

$k = m$	$\lvert S \rvert$	$\lvert V \rvert$	GOA [LeMa96b] + post-pass	GA	%
2	50	10	5.9	3.6	61
2	50	20	11.4	6.5	57
2	50	40	9.1	6.9	67
2	100	10	12.6	4.0	32
2	100	50	38.6	28.7	74
2	100	90	11.0	9.5	86
4	50	10	5.0	4.0	80
4	50	20	7.6	5.1	67
4	50	40	9.0	6.2	69
4	100	10	5.2	4.1	79
4	100	50	17.3	9.3	54
4	100	90	8.7	7.2	83
8	50	10	5.0	4.0	80
8	50	20	8.9	5.2	58
8	50	40	11.7	7.3	62
8	100	10	5.0	4.1	82
8	100	50	15.0	9.9	66
8	100	90	12.2	8.6	70
				average	68

Table A.3.   Performance of the GA for $(k, 0, 2)$-OA and $(k, 4, 0)$-OA (chapter 2). Columns 3 and 4 give the results for [WeGo97a] and the $(k, 0, 2)$-OA algorithm, while column 5 gives the percentage of the GA. Columns 6 and 7 show the results of the GA when using modify registers dynamically.

$(\lvert V \rvert, \lvert S \rvert)$	k	[WeGo97a] $l = 2$ $m = 0$	GA $l = 2$ $m = 0$	%	GA $l = 0$ $m = 4$	%
(15,20)	1	1.8	2.2	122	0.9	50
(10,50)	1	11.9	14.1	118	10.2	86
(10,50)	2	1.8	1.2	67	1.0	56
(25,50)	2	6.7	7.6	113	6.7	100
(40,50)	4	3.0	1.1	37	0.8	27
(50,100)	4	15.6	17.4	112	13.9	89
average				95		68

Table A.4.   Experimental evaluation of the three AR allocation algorithms from chapter 2 for different problem parameters: n is the access sequence length, D is the maximum offset distance of array accesses, and r is the auto-increment range. Column 2 gives the average number of ARs needed by the matching-based algorithm, while column 3 shows the average number of extra instructions for address computation at the end of a loop iteration. Column 4 lists the number of ARs heuristically found by the path-based algorithm, and column 5 shows the optimal results computed by the branch-and-bound algorithm. For the latter, the CPU times are shown in column 6. Finally, column 7 gives the percentage of overhead of the path-based heuristic as compared to the optimal solutions. Variation of parameter r (third section of the table) shows that the CPU time is inversely related to r. This is explained by the fact, that also the average number of extra address computations incurred by the matching-based algorithm is inversely related to r. This means, that for large r the matching-based algorithm yields zero-cost (and thus optimal) solutions in most cases, so that the branch-and-bound procedure can terminate immediately. With respect to parameter D, the CPU time assumes a maximum for $D = 8$. This can be explained as the result of two contrary effects: The CPU time for branch-and-bound grows with the average number of extra address computations incurred by the matching-based algorithm. As can be seen in column 3, this number grows with D, so that in turn the CPU time tends to grow with D. Simultaneously, larger D values imply less edges in the distance graph and thus lower CPU times. This effect obviously becomes predominant for $D > 8$.

parameter	# ARs matching	# unit-cost address comp.	# ARs path-based	# ARs B&B	CPU sec B&B	overhead path-based
$n = 5$	2.05	0.18	2.18	2.15	< 0.01	1.39 %
$n = 10$	2.65	0.36	2.87	2.79	0.01	3.19 %
$n = 15$	2.94	0.54	3.27	3.11	0.13	5.10 %
$n = 20$	3.18	0.66	3.58	3.36	2.63	6.84 %
$n = 25$	3.29	0.79	3.81	3.49	31.01	9.06 %
$D = 4$	1.54	0.21	1.65	1.56	5.48	5.37 %
$D = 8$	2.07	0.39	2.30	2.17	14.78	5.94 %
$D = 16$	3.05	0.61	3.44	3.24	4.45	6.23 %
$D = 32$	4.63	0.81	5.18	4.94	2.30	4.93 %
$r = 1$	5.11	1.09	5.75	5.63	15.89	2.18 %
$r = 3$	3.08	0.53	3.41	3.17	10.29	7.61 %
$r = 7$	1.84	0.28	2.05	1.85	0.73	10.49 %
$r = 15$	1.25	0.12	1.36	1.26	0.11	7.91 %
average						5.86 %

Table A.5.   Experimental results for CSE register allocation (chapter 3). Column 1 mentions the source of tested basic blocks with common subexpressions. Columns 2 and 3 provide the number of CSEs and their uses, respectively. Column 4 gives the number of CSEs that have been assigned to special-purpose registers by the proposed algorithm. Columns 5 and 6 show the percentage of costs (code size) and memory accesses for CSEs as compared to a non-optimized solution with all CSEs being mapped to memory. Finally, column 7 gives the required CPU time in seconds.

source	CSEs	CSE uses	reg CSEs	cost (%)	mem (%)	CPU
IIR filter 1	2	5	1	91	57	1
IIR filter n	12	26	3	89	72	10
FFT	13	32	2	97	95	7
2-dim FIR filter	3	6	3	94	42	2
ADPCM	18	36	12	93	31	19
LMS filter	2	5	0	100	100	1
n complex updates	12	26	2	96	89	7
basic functions	2	4	2	76	40	3
receiver	52	113	17	95	73	34
main buffer ctrl	3	8	3	97	10	4
compr preproc ctrl	14	35	12	97	44	16
IDCT	41	100	5	97	85	79
downsampling	37	81	17	92	66	32
transcoding compr	4	8	2	85	67	4
IDCT	49	99	2	100	97	85
motion vector dec	24	58	7	94	80	20
DCT block decoding	24	54	18	83	59	41
motion comp predict	48	127	13	100	89	38
quantization	11	22	11	76	47	12
grey scale coding	15	31	8	93	74	33
bitstream functions	3	7	2	100	83	4
DCT coeff quant	18	38	12	92	64	18
average				93	67	

Table A.6. Experimental results of code selection with SIMD instructions (chapter 5). The upper half of the table refers to TI C6201, while the lower half gives results for the Philips Trimedia. Column 1 mentions the source code, while column 2 gives the data type (short = 16 bits, char = 8 bits). Column 3 mentions the number of duplications of the loop body so as to reveal enough instruction-level parallelism. Columns 4 and 5 give the number of instructions generated without and with the use of SIMD instructions enabled. Columns 6 and 7 show the number of variables and constraints in the corresponding ILP models. Finally, column 8 mentions the CPU time.

source	data type	unrl.	no SIMD	SIMD	var	cons	CPU sec
			TI C6201				
vector add	short	1	8	4	524	662	0.7
IIR filter	short	0	21	17	1078	1359	2.9
convolution	short	1	8	6	467	620	0.6
FIR filter	short	1	15	11	568	710	0.9
N complex updates	short	1	20	16	1004	1260	3.0
image compositing	short	1	14	11	998	1112	3.1
			Trimedia TM1000				
vector add	short	1	8	4	536	660	0.7
IIR filter	short	0	22	22	1084	1188	5.1
convolution	short	1	8	8	638	783	0.9
FIR filter	short	1	15	9	604	722	0.9
N complex updates	short	1	20	20	1059	1298	4.7
image compositing	short	1	14	7	1018	1243	3.2
vector add	char	3	16	4	1049	1397	5.0
FIR filter	char	3	36	18	1129	1369	26.5

Table A.7.   Characteristics of C source codes used for evaluating the ITE optimization from chapter 6. Column 2 gives the number of ITE statements, column 3 gives the maximum ITE nesting level, and column 4 mentions the code size in terms of IR statements.

source	# ITE	nest	size
adapt_quant	4	3	16
adapt_predict1	3	1	29
adapt_predict2	6	2	44
diff_comp	2	1	22
outp_conv	4	2	34
code_adj1	5	5	19
code_adj2	17	9	86
code_adj3	17	5	95
detect_pos	7	3	45
find_mv	4	4	45

Table A.8.   Worst-case execution time (instruction cycles) for the C codes from table A.7 for a TI C6201 VLIW DSP. Column 2 gives the number of cycles when only using the C-JMP scheme. Likewise, column 3 corresponds to solutions only using the C-INS scheme. Column 4 gives the results for the dynamic programming based optimization algorithm, while column 5 provides the results obtained with the TI C6201 C compiler.

source	C-JMP	C-INS	opt	TI
adapt_quant	21	11	11	15
adapt_predict1	12	13	13	13
adapt_predict2	26	21	22	27
diff_comp	9	12	12	10
outp_conv	26	30	24	21
code_adj1	32	23	23	30
code_adj2	57	173	49	51
code_adj3	39	244	30	41
detect_pos	28	27	27	29
find_mv	27	30	30	28

Table A.9. Code size (in instruction words) for the C codes from table A.7 for a TI C6201 VLIW DSP. The column structure corresponds to that of table A.8.

source	C-JMP	C-INS	opt	TI
adapt_quant	14	23	23	13
adapt_predict1	13	14	14	18
adapt_predict2	32	43	30	27
diff_comp	9	12	12	12
outp_conv	22	50	34	23
code_adj1	21	42	29	21
code_adj2	59	253	75	56
code_adj3	99	305	109	87
detect_pos	31	44	44	32
find_mv	40	57	57	38

Table A.10. Detailed results for function inlining application study from chapter 7 (GSM encoder on a TI C6201). The highlighted line denotes the absolute minimum in the number of execution cycles.

limit %	calls abs	calls %	est size	real size	err %	cycles abs	cycles %
100	10,292,056	100	–	67,820	–	27,400,547	100
105	7,618,479	74	71,200	70,284	1	24,095,022	88
110	5,893,530	57	74,536	72,876	2	19,560,628	71
115	4,984,329	48	77,976	77,796	1	20,190,858	74
120	4,403,360	43	81,372	80,772	1	20,518,980	75
125	3,892,613	38	84,768	82,636	3	18,235,114	67
130	3,427,558	33	88,148	87,908	1	18,527,926	68
135	2,414,683	23	91,544	89,796	2	18,416,065	67
140	2,385,409	23	93,812	91,940	2	18,750,981	68
145	1,872,297	18	98,320	97,956	1	18,796,095	69
150	1,797,790	17	101,716	100,484	1	19,136,175	70

Appendix B
Example for the LANCE V2.0 IR

C source code for the FIR filter algorithm from the DSPStone benchmark suite [ZVSM94].

```c
#define STORAGE_CLASS register
#define TYPE   int
#define LENGTH 16

void pin_down(TYPE * px, TYPE * ph, TYPE y)
{
  STORAGE_CLASS TYPE i;
  for (i = 1; i <= LENGTH; i++) {
      *px++ = i;
      *ph++ = i;
    }

}

TYPE main()
{
  static TYPE  x[LENGTH];
  static TYPE  h[LENGTH];
  static TYPE  x0 = 100;
  STORAGE_CLASS TYPE i ;
  STORAGE_CLASS TYPE *px, *px2 ;
  STORAGE_CLASS TYPE *ph ;
  STORAGE_CLASS TYPE y;

  pin_down(x, h, y);
  ph  = &h[LENGTH-1] ;
  px  = &x[LENGTH-1]  ;
  px2 = &x[LENGTH-2]  ;

  y = 0;

  for (i = 0; i < LENGTH - 1; i++) {
      y += *ph-- * *px ;
      *px-- = *px2-- ;
    }

  y += *ph * *px ;
  *px = x0 ;
  pin_down(x, h, y);
  return ((TYPE) y);
}
```

191

Unoptimized IR-C code for the FIR example, generated for a SUN workstation host. Auxiliary variables inserted by the frontend have a "t" prefix. All local variables get unique numbers as suffixes, and local static variables are transformed into global variables. Blocks of IR statements are annotated with source code information for debugging purposes.

```c
void pin_down(int *,int *,int );
int main();

static int _local_static_x_8[16];

static int _local_static_h_9[16];

static int _local_static_x0_10 = 100;

void pin_down(int *px_2,int *ph_3,int y_4)
{
 register int i_6;
 int t1;
 int t2;
 int t3;
 int t12;
 int *t4;
 char *t5;
 char *t6;
 int *t7;
 int *t8;
 char *t9;
 char *t10;
 int *t11;


 /* 10 "fir.c" */
 /* $ for (i = 1; i <= 16 ; i++)  $ */

        i_6 = 1;
        LL3:
        t1 = i_6 <= 16;
        t12 = !t1;
        if (t12) goto LL1;


 /* 12 "fir.c" */
 /* $ *px++ = i; $ */

        t4 = px_2;
        t6 = (char *)t4;
        t5 = t6 + 4;
        t7 = (int *)t5;
```

```
        px_2 = t7;
        *t4 = i_6;

/* 13 "fir.c" */
/* $ *ph++ = i; $ */

        t8 = ph_3;
        t10 = (char *)t8;
        t9 = t10 + 4;
        t11 = (int *)t9;
        ph_3 = t11;
        *t8 = i_6;

/* 10 "fir.c" */
/* $ for (i = 1; i <= 16 ; i++)   $ */

        LL2:
        t2 = i_6;
        t3 = t2 + 1;
        i_6 = t3;
        goto LL3;

        LL1:
        return;
}

int main()
{
  register int i_11;
  register int *px_12;
  register int *px2_13;
  register int *ph_14;
  register int y_15;
  int t14;
  char *t15;
  int t16;
  char *t17;
  int *t18;
  int t19;
  char *t20;
  int t21;
  char *t22;
  int *t23;
  int t24;
  char *t25;
  int t26;
  char *t27;
  int *t28;
  int t29;
  int t30;
```

```
int t31;
int t32;
int t47;
int t48;
int t49;
int *t50;
int *t33;
char *t34;
char *t35;
int *t36;
int t37;
int t38;
int *t39;
char *t40;
char *t41;
int *t42;
int *t43;
char *t44;
char *t45;
int *t46;


/* 30 "fir.c" */
/* $ pin_down(x, h, y); $ */

        pin_down(_local_static_x_8,_local_static_h_9,y_15);

/* 32 "fir.c" */
/* $ ph  = &h[16 -1] ;  $ */

        t14 = 16 - 1;
        t17 = (char *)_local_static_h_9;
        t16 = t14 * 4;
        t15 = t17 + t16;
        t18 = (int *)t15;
        ph_14 = t18;

/* 33 "fir.c" */
/* $ px  = &x[16 -1] ;  $ */

        t19 = 16 - 1;
        t22 = (char *)_local_static_x_8;
        t21 = t19 * 4;
        t20 = t22 + t21;
        t23 = (int *)t20;
        px_12 = t23;

/* 34 "fir.c" */
/* $ px2 = &x[16 -2] ;  $ */
```

```
        t24 = 16 - 2;
        t27 = (char *)_local_static_x_8;
        t26 = t24 * 4;
        t25 = t27 + t26;
        t28 = (int *)t25;
        px2_13 = t28;

/* 37 "fir.c" */
/* $ y = 0; $ */

        y_15 = 0;

/* 39 "fir.c" */
/* $ for (i = 0; i < 16  - 1; i++) $ */

        i_11 = 0;
        LL6:
        t29 = 16 - 1;
        t30 = i_11 < t29;
        t47 = !t30;
        if (t47) goto LL4;

/* 41 "fir.c" */
/* $ y += *ph-- * *px ;  $ */

        t33 = ph_14;
        t35 = (char *)t33;
        t34 = t35 - 4;
        t36 = (int *)t34;
        ph_14 = t36;
        t37 = *t33 * *px_12;
        t38 = y_15 + t37;
        y_15 = t38;

/* 42 "fir.c" */
/* $ *px-- = *px2-- ;  $ */

        t39 = px2_13;
        t41 = (char *)t39;
        t40 = t41 - 4;
        t42 = (int *)t40;
        px2_13 = t42;
        t43 = px_12;
        t45 = (char *)t43;
        t44 = t45 - 4;
        t46 = (int *)t44;
        px_12 = t46;
        *t43 = *t39;
```

```
/* 39 "fir.c" */
/* $ for (i = 0; i < 16  - 1; i++) $ */

        LL5:
        t31 = i_11;
        t32 = t31 + 1;
        i_11 = t32;
        goto LL6;

        LL4:

/* 45 "fir.c" */
/* $ y += *ph * *px ; $ */

        t48 = *ph_14 * *px_12;
        t49 = y_15 + t48;
        y_15 = t49;

/* 46 "fir.c" */
/* $ *px = x0 ;   $ */

        t50 = &_local_static_x0_10;
        *px_12 = *t50;

/* 49 "fir.c" */
/* $ pin_down(x, h, y); $ */

        pin_down(_local_static_x_8,_local_static_h_9,y_15);

/* 51 "fir.c" */
/* $ return ((int ) y);  $ */

        return y_15;

}
```

References

[ACE00]   ACE Associated Compiler Experts: www.ace.nl, 2000

[AGT89]   A.V. Aho, M. Ganapathi, S.W.K Tjiang: *Code Generation Using Tree Matching and Dynamic Programming*, ACM Trans. on Programming Languages and Systems 11, No. 4, 1989

[AhJo76]   A.V. Aho, S.C. Johnson: *Optimal Code Generation for Expression Trees*, Journal of the ACM, vol. 23, no. 3, 1976

[AhUl72]   A.V. Aho, J.D. Ullman: *The Theory of Parsing, Translation and Compiling*, vols. I and II, Prentice Hall, 1972

[AHM97]   D.I. August, W.W. Hwu, S.A. Mahlke: *A Framework for Balancing Control Flow and Predication*, 30th Int. Symp. on Microprogramming and Microarchitecture (MICRO-97), 1997

[AiNi88]   A. Aiken, A. Nicolau: *A Development Environment for Horizontal Microcode*, IEEE Trans. on Software Engineering, no. 14, 1988

[AJU77]   A.V. Aho, S.C. Johnson, J.D. Ullman: *Code Generation for Expressions with Common Subexpressions*, Journal of the ACM, vol. 24, no. 1, 1977

[AKPW83]   J.R. Allen, K. Kennedy, C. Porterfield, J. Warren: *Conversion of Control Dependence into Data Dependence*, 10th ACM Symp. on Principles of Programming Languages (POPL), 1983

[AlMa95]   M. Alt, F. Martin: *Generation of Efficient Interprocedural Analyzers with PAG*, in: A. Mycroft (ed.): *Static Analysis*, LNCS 983. Springer, 1995

[AML96]   G. Araujo, S. Malik, M. Lee: *Using Register Transfer Paths in Code Generation for Heterogeneous Memory-Register Architectures*, 33rd Design Automation Conference (DAC), 1996

[AMS00]   Austria Mikro Systeme International: asic.amsint.com/databooks/digital/gepard.html, 2000

[Appe98]   A.W. Appel: *Modern Compiler Implementation in C*, Cambridge University Press, 1998

[Arch00]   Archelon Inc.: www.archelon.com, 2000

[ARM00]   ARM home page: www.arm.com, 2000

[ArMa95]   G. Araujo, S. Malik: *Optimal Code Generation for Embedded Memory Non-Homogeneous Register Architectures*, 8th Int. Symp. on System Synthesis (ISSS), 1995

[Arno00]   G. Arnout: *SystemC Standard*, Asia South Pacific Design Automation Conference (ASP-DAC), 2000

[ASM96]   G. Araujo, A. Sudarsanam, S. Malik: *Instruction Set Design and Optimizations for Address Computation in DSP Architectures*, 9th Int. Symp. on System Synthesis (ISSS), 1996

[ASU86]   A.V. Aho, R. Sethi, J.D. Ullman: *Compilers - Principles, Techniques, and Tools*, Addison-Wesley, 1986

[Assm98]   U. Assmann: *OPTIMIX, A Tool for Rewriting and Optimizing Programs*, Graph Grammar Handbook, Chapman-Hall, 1998

[BaLe99a]   S. Bashford, R. Leupers: *Phase-Coupled Mapping of Data Flow Graphs to Irregular Data Paths*, Design Automation for Embedded Systems, Vol. 4, No. 2/3, Kluwer Academic Publishers, 1999

[BaLe99b]   S. Bashford, R. Leupers: *Constraint Driven Code Selection for Fixed-Point DSPs*, 36th Design Automation Conference (DAC), 1999

[Bars99]   T. Barschdorf: *Codegenerierung für den digitalen Signalprozessor TI TMS320C5x / Code Generation for the digital signal processor TI TMS320C5x*, Diploma thesis (in German), University of Dortmund, Dept. of Computer Science, 1999

[Bart92]   D.H. Bartley: *Optimizing Stack Frame Accesses for Processors with Restricted Addressing Modes*, Software – Practice and Experience, vol. 22(2), 1992

[BDB90]   A. Balachandran, D.M. Dhamdere, S. Biswas: *Efficient Retargetable Code Generation Using Bottom-Up Tree Pattern Matching*, Comput. Lang. vol. 15, no. 3, 1990

[BeDe98]   L. Benini, G. De Micheli: *Dynamic Power Management - Design Techniques and CAD Tools*, Kluwer Academic Publishers, 1998

[BEH91]   D. Bradlee, S. Eggers, R. Henry: *Integrating Register Allocation and Instruction Scheduling for RISCs*, 4th Int. Conf. on Architectural Support for Programming Languages and Operating Systems (ASPLOS), 1991

[Bela66]   L.A. Belady: *A Study of Replacement Algorithms for a Virtual-Storage Computer*, IBM System Journals 5(2), 1966

[Bisc92]   K.M. Bischoff: *Design, Implementation, Use, and Evaluation of Ox: An Attribute-Grammer Compiling System based on Yacc, Lex, and C*, Technical Report 92-31, Dept. of Computer Science, Iowa State University, 1992

[BJER98]   R. Bhargava, L.K. John, B.L. Evans, R. Radhakrishnan: *Evaluating MMX Technology Using DSP and Multimedia Applications*, 31st Int. Symp. on Microprogramming and Microarchitecture (MICRO-98), 1998

[BLM98]   A. Basu, R. Leupers, P. Marwedel: *Register-Constrained Address Computation in DSP Programs*, Design Automation & Test in Europe (DATE), 1998

[BML96]   S.S. Bhattacharyya, P.K. Murthy, E.A. Lee: *Software Synthesis from Dataflow Graphs*, Kluwer Academic Publishers, 1996

[BoGi77]   F.T. Boesch, J.F. Gimpel: *Covering the Points of a Digraph with Point-Disjoint Paths and Its Application to Code Optimization*, Journal of the ACM, vol. 24, no. 2, 1977

[Brig92]   P. Briggs: *Register Allocation via Graph Coloring*, Doctoral thesis, Dept. of Computer Science, Rice University, Houston/Texas, 1992

[BrSe76]   J. Bruno, R. Sethi: *Code Generation for a One-Register Machine*, Journal of the ACM, no. 23, 1976

[BSBC95]   T.S. Brasier, P.H. Sweany, S. Carr, S.J. Beaty: *CRAIG: A Practical Framework for Combining Instruction Scheduling and Register Allocation*, Int. Conf. on Parallel Architectures and Compilation Techniques (PACT), 1995

[CAP99]   P. Centoducatte, G. Araujo, R. Pannain: *Compressed Code Execution on DSP Architectures*, 12th Int. System Synthesis Symposium (ISSS), 1999

[CASE99]   The Second International Workshop on Compiler and Architecture Support for Embedded Systems (CASES'99), www.capsl.udel.edu/CONFERENCES/CASES99, 1999

[CDN94]   A. Capitanio, N. Dutt, A. Nicolau: *Partitioning of Variables for Multiple-Register-File Architectures via Hypergraph Coloring*, in: M. Cosnard, G.R. Gao, G.M. Silberman (eds.): IFIP Trans. A-50 – Parallel Architectures and Compilation Techniques, North Holland, 1994

[Chai82]   G.J. Chaitin: *Register Allocation and Spilling via Graph Coloring*, SIGPLAN Symp. on Compiler Construction, 1982

[ChLi98]   W. Chen, Y. Lin: *Addressing Optimization for Loop Execution Targeting DSP with Auto-Increment/Decrement Architecture*, 11th Int. System Synthesis Symposium (ISSS), 1999

[ChLi99]   W. Chen, Y. Lin: *Code Generation of Nested Loops for DSP Processors with Heterogeneous Registers and Structural Pipelining*, ACM Trans. on Design Automation of Electronic Systems (TO-DAES), Vol. 4, No. 3, 1999

[CHT92]   K.D. Cooper, M.W. Hall, L. Torczon: *Unexpected Side Effects of Inline Substitution*, ACM Letters on Programming Languages and Systems, vol. 1, no. 1, 1992

[CoCo00]   CoCoLab: www.cocolab.de, 2000

[CWKL+99]   M. Coors, O. Wahlen, H. Keding, O. Lüthje, H. Meyr: *TI C62x Performance Code Optimization*, DSP Deutschland, 2000

[Davi91]   L. Davis: *Handbook of Genetic Algorithms*, Van Nostrand Reinhold, 1991

[Davi98]   F. David: *Optimierte Adresszuweisung in DSP-Compilern / Optimized Address Assignment in DSP Compilers*, Diploma thesis (in German), University of Dortmund, Dept. of Computer Science, 1998

[Dege95]   J. Degener: *ANCI C Yacc Grammar*, www.lysator.liu.se/c/ANSI-C-grammar-y.html, 1995

[DLSM81]   S. Davidson, D. Landskov, B.D. Shriver, P.W. Mallett: *Some Experiments in Local Microcode Compaction for Horizontal Machines*, IEEE Trans. on Computers, Vol. 30, No. 7, 1981

[Drec98]   R. Drechsler: *Evolutionary Algorithms for VLSI CAD*, Kluwer Academic Publishers, 1998

[DrHa89]   D. Drusinsky, D. Harel: *Using Statecharts for Hardware Description and Design*, IEEE Trans. on Computer-Aided Design, Vol. 8, No. 7, 1989

[EcKr99]   E. Eckstein, A. Krall: *Minimizing Cost of Local Variables Access for DSP Processors*, ACM Workshop on Languages, Compilers, and Tools for Embedded Systems (LCTES), 1999

[EDG00]   Edison Design Group: www.edg.com, 2000

[EHB93]   R. Ernst, J. Henkel, T. Benner: *Hardware-Software Cosynthesis for Microcontrollers*, IEEE Design & Test Magazine, No. 12, 1993

[Eind00]   Eindhoven University of Technology: ftp.es.ele.tue.nl/pub/lp_solve/, 2000

[Ertl99]   M.A. Ertl: *Optimal Code Selection in DAGs*, ACM Symp. on Principles of Programming Languages (POPL), 1999

[ESL89]   H. Emmelmann, F.W. Schröer, R. Landwehr: *BEG – A Generator for Efficient Backends*, ACM SIGPLAN Conference on Programming Language Design and Implementation (PLDI), SIGPLAN Notices 24, no. 7, 1989

[FaKn93]   A. Fauth, A. Knoll: *Translating Signal Flowcharts into Microcode for Custom Digital Signal Processors*, Int. Conf. on Signal Processing (ICSP), 1993

[FDF98a]   P. Faraboschi, G. Desoli, J.A. Fisher: *VLIW Architectures for DSP and Multimedia Applications – The Latest Word in Digital and Media Processing*, IEEE Signal Processing Magazine, March 1998

[FDF98b]   P. Faraboschi, G. Desoli, J.A. Fisher: *Clustered Instruction-Level Parallel Processors*, Technical Report HPL-98-204, HP Labs, USA, 1998

[FHP92a]   C.W. Fraser, D.R. Hanson, T.A. Proebsting: *Engineering a Simple, Efficient Code Generator Generator*, ACM Letters on Programming Languages and Systems, vol. 1, no. 3, 1992

[FHP92b]   C.W. Fraser, R.R. Henry, T.A. Proebsting: *BURG – Fast Optimal Instruction Selection and Tree Parsing*, ACM SIGPLAN Notices 27 (4), 1992

[FiDi98]   R.J. Fisher, H.G. Dietz: *Compiling for SIMD Within a Register*, 11th Annual Workshop on Languages and Compilers for Parallel Computing (LCPC98), 1998

[FiDi99]   R.J. Fisher, H.G. Dietz: *The Scc Compiler: SWARing at MMX and 3DNow!*, 12th Annual Workshop on Languages and Compilers for Parallel Computing (LCPC99), 1999

[Fish81]   J.A. Fisher: *Trace Scheduling: A Technique for Global Microcode Compaction*, IEEE Trans. on Computers, vol. 30, no. 7, 1981

[FMW97]   C. Ferdinand, F. Martin, R. Wilhelm: *Applying Compiler Techniques to Cache Behavior Prediction*, ACM SIGPLAN Workshop on Languages, Compilers and Tools for Real-Time System (LCTRTS), 1997

[FrHa95]   C. Fraser, D. Hanson: *A Retargetable C Compiler: Design And Implementation*, Addison-Wesley, 1995, www.cs.princeton.edu/software/lcc

[FSW94]   C. Ferdinand, H. Seidl, R. Wilhelm: *Tree Automata for Code Selection*, Acta Informatica 31, Springer Verlag 1994

[FVM95]   A. Fauth, J. Van Praet, M. Freericks: *Describing Instruction-Set Processors in nML*, European Design and Test Conference (ED & TC), 1995

[FWD+98]   G. Fettweis, M. Weiss, W. Drescher et al.: *Breaking New Grounds Over 3000 M MAC/s: A Broadband Mobile Multimedia Modem DSP*, DSP Deutschland, 1998

[GaJo79]   M.R. Gary, D.S. Johnson: *Computers and Intractability – A Guide to the Theory of NP-Completeness*, Freemann, 1979

[GBF97]   R. Gupta, D.A. Berson, J.Z. Fang: *Path Profile Guided Partial Dead Code Elimination Using Predication*, Int. Conf. on Parallel Architectures and Compilation Techniques (PACT), 1997

[GDWL92]   D. Gajski, N. Dutt, A. Wu, S. Lin: *High-Level Synthesis – Introduction to Chip and System Design*, Kluwer Academic Publishers, 1992

[Gebo97a] C. Gebotys: *DSP Address Optimization Using a Minimum Cost Circulation Technique*, Int. Conference on Computer-Aided Design (ICCAD), 1997

[Gebo97b] C. Gebotys: *Low Energy Memory and Register Allocation Using Network Flow*, 34th Design Automation Conference (DAC), 1997

[GeEl92]   C. Gebotys, M. Elmasry: *Optimal VLSI Architectural Synthesis*, Kluwer Academic Publishers, 1992

[GGMC00]   S. Gupta, R. Gupta, M. Miranda, F. Catthoor: *Analysis of High-Level Address Code Transformations for Programmable Processors*, Design Automation & Test in Europe (DATE), 2000

[GHZG99]   T. Gaul, A. Heberle, W. Zimmermann, W. Goerigk: *Construction of Verified Software Systems with Program-Checking: An Application To Compiler Back-Ends*, Proc. Workshop on Runtime Result Verification (RTRV), 1999

[Glan77]   R.S. Glanville: *A Machine Independent Algorithm for Code Generation and its Use in Retargetable Compilers*, Doctoral thesis, University of California at Berkeley, 1977

[GNR00]   N. Ghazal, R. Newton, J. Rabaey: *Retargetable Estimation Scheme for DSP Architecture Selection*, Asia South Pacific Design Automation Conference (ASP-DAC), 2000

[GNU00]   Free Software Foundation: www.gnu.org, 2000

[GoHs88]   J. Goodman, W. Hsu: *Code Scheduling and Register Allocation in Large Basic Blocks*, ACM SIGPLAN Conference on Programming Language Design and Implementation (PLDI), 1988

[GuLu95]   H. Gunnarson, T. Lundqvist: *Porting the GNU C Compiler to the Thor Microprocessor*, Master Thesis, Document No. TOR/TNT/0028/SE, www.ce.chalmers.se/~thomasl/publications/thesis95.html, Saab Ericsson Space AB, 1995

[GVD89]   G. Goossens, J. Vandewalle, H. De Man: *Loop Optimization in Register-Transfer Scheduling for DSP Systems*, 26th Design Automation Conference (DAC), 1989

[GVNG94]   D. Gajski, F. Vahid, S. Narayan, J. Gong: *Specification and Design of Embedded Systems*, Prentice Hall, 1994

[GZD+00]   D. Gajski, J. Zhu, R. Dömer, A. Gerstlauer, S. Zhao: *SpecC: Specification Language and Methodology*, Kluwer Academic Publishers, 2000

[HaDe98]   S. Hanono, S. Devadas: *Instruction Selection, Resource Allocation, and Scheduling in the AVIV Retargetable Code Generator*, 35th Design Automation Conference (DAC), 1998

[Hare87]   D. Harel: *Statecharts: A Visual Formalism for Complex Systems*, Science of Computer Programming 8, North-Holland, 1987

[Hart92]   R. Hartmann: *Combined Scheduling and Data Routing for Programmable ASIC Systems*, European Conference on Design Automation (EDAC), 1992

[HePa90]   J.L. Hennessy, D.A. Patterson: *Computer Architecture – A Quantitative Approach*, Morgan Kaufmann Publishers Inc., 1990

[HHD97]   G. Hadjiyiannis, S. Hanono, S. Devadas: *ISDL: An Instruction-Set Description Language for Retargetability* 34th Design Automation Conference (DAC), 1997

[Hilf85]   P. Hilfinger: *A High-Level Language and Silicon Compiler for Digital Signal Processing*, Custom Integrated Circuits Conference (CICC), 1985

[HoSa87]   E. Horowitz, S. Sahni: *Fundamentals of Data Structures in PASCAL*, 2nd Edition, Computer Science Press Inc., 1987

[HRR+97]   M. Hartoog, J. Rowson, P. Reddy, et al.: *Generation of Software Tools from Processor Descriptions for Hardware/Software Codesign*, 34th Design Automation Conference (DAC), 1997

[Hwu98]   W. Hwu: *Introduction to Predicated Execution*, IEEE Computer, Jan 1998

[IBM00]   IBM: OSL home page, www6.software.ibm.com/es/oslv2/features/welcome.htm, 2000

[ICD00]   Informatik Centrum Dortmund: www.icd.de, 2000

[Infi00]   Infineon Technologies: *Carmel – The DSP Core*, product infor-
             mation, www.infineon.com/us/carmel, 2000

[Inte00a]   Intel: *MMX Technology Application Notes*,
             developer.intel.com/drg/mmx/appnotes, 2000

[Inte00b]   Intel: *Coding Techniques for the Streaming SIMD
             Extensions With the Intel C/C++ Compiler*, devel-
             oper.intel.com/vtune/newsletr/methods.htm, 2000

[JaVe99]   M.F. Jacome, G. de Veciana: *Lower Bound on Latency for VLIW
             ASIP Data Paths*, Int. Conf. on Computer-Aided Design (ICCAD),
             1999

[JoAl90]   R.B. Jones, V.H. Allan: *Software Pipelining: A Comparison and
             Improvement*, 22nd Annual Workshop on Microprogramming and
             Microarchitecture (MICRO-23), 1990

[JPEG00]   Independent JPEG group: www.ijg.org, 2000

[KAJW96]   S. Kumar, J.H. Aylor, B.W. Johnson, W.A. Wulf: *The Codesign of
             Embedded Systems*, Kluwer Academic Publishers, 1996

[KaLa98]   D. Kästner, M. Langenbach: *Integer Linear Programming vs.
             Graph-Based Methods in Code Generation*, Technical Report
             A/01/98, Dept. of Computer Science, University of Saarland, Ger-
             many, 1998

[KeLi70]   B.W. Kernighan, S. Lin: *An Efficient Heuristic Procedure for Par-
             titioning Graphs*, Bell Sys. Tech. Journal, Vol. 49, 1970

[KeRi88]   B.W. Kernighan, D.M. Ritchie: *The C Programming Language*,
             2nd Edition, Prentice Hall, 1988

[KGV83]   S. Kirkpatrick, C.D. Gelatt, M.P. Vecchi: *Optimzation by Simu-
             lated Annealing*, Science, Vol. 220, 1983

[KMTP+95]   L. Kohn, G. Maturana, M. Tremblay, A. Prabhu, G. Zyner: *The
             Visual Instruction Set (VIS) in Ultra-SPARC*, Proc. Compcon '95,
             1995

[KNDK95]   D.J. Kolson, A. Nicolau, N. Dutt, K. Kennedy: *Optimal Register
             Assignment to Loops for Embedded Code Generation*, 8th Int.
             Symp. on System Synthesis (ISSS), 1995

[KRS92]   J. Knoop, O. Rüthing, B. Steffen: *Lazy Code Motion*, ACM SIG-
             PLAN Conference on Programming Language Design and Imple-
             mentation (PLDI), 1992

[KRS94]   J. Knoop, O. Rüthing, B. Steffen: *Partial Dead Code Elimination*,
             ACM SIGPLAN Conference on Programming Language Design
             and Implementation (PLDI), 1994

[KSN97]   N. Kogure, N. Sugino, A. Nishihara: *Memory Address Alloca-tion Method for a DSP with ± 2 Update Operations in Indirect Addressing*, European Conference on Circuit Theory and Design (ECCTD), 1997

[KuRo00]   T. Kuhn, W. Rosenstiel: *Java Based Object Oriented Hardware Specification and Synthesis*, Asia South Pacific Design Automation Conference (ASP-DAC), 2000

[LaCe93]   M. Langevin, E. Cerny: *An Automata-Theoretic Approach to Local Microcode Generation*, European Conference on Design Automation (EDAC), 1993

[Lam88]   M. Lam: *Software Pipelining: An Effective Scheduling Technique for VLIW machines*, ACM SIGPLAN Conference on Programming Language Design and Implementation (PLDI), 1988

[LaMa97]   B. Landwehr, P. Marwedel: *A New Optimization Technique for Improving Resource Exploitation and Critical Path Minimization*, 10th Int. Symp. on System Synthesis (ISSS), 1997

[LANC00]   LANCE Software: University of Dortmund, Dept. of Computer Science 12, ls12-www.cs.uni-dortmund.de/~leupers, 2000

[LBM98]   R. Leupers, A. Basu, P. Marwedel: *Optimized Array Index Computation in DSP Programs*, Asia South Pacific Design Automation Conference (ASP-DAC), 1998

[LCGD94]   D. Lanneer, M. Cornero, G. Goossens, H. De Man: *Data Routing: A Paradigm for Efficient Data-Path Synthesis and Code Generation*, 7th Int. Symp. on High-Level Synthesis (HLSS), 1994

[LCT99]   ACM SIGPLAN 1999 Workshop on Languages, Compilers, and Tools for Embedded Systems (LCTES'99), www.cs.indiana.edu/~liu/lctes99, 1999

[LDK+95a]   S. Liao, S. Devadas, K. Keutzer, S. Tjiang, A. Wang: *Code Optimization Techniques for Embedded DSP Microprocessors*, 32nd Design Automation Conference (DAC), 1995

[LDK+95b]   S. Liao, S. Devadas, K. Keutzer, S. Tjiang, A. Wang: *Storage Assignment to Decrease Code Size*, ACM SIGPLAN Conference on Programming Language Design and Implementation (PLDI), 1995

[LDK+95c]   S. Liao, S. Devadas, K. Keutzer, S. Tjiang: *Instruction Selection Using Binate Covering for Code Size Optimization*, Int. Conf. on Computer-Aided Design (ICCAD), 1995

[LDK99]   S. Liao, S. Devadas, K. Keutzer: *A Text-Compression Based Method for Code Size Minimization in Embedded Systems*, ACM

Trans. on Design Automation of Electronic Systems (TODAES), Vol. 4, No. 1, 1999

[LeDa98]   R. Leupers, F. David: *A Uniform Optimization Technique for Offset Assignment Problems*, 11th Int. System Synthesis Symposium (ISSS), 1998

[Lee88]   E.A. Lee: *Programmable DSP Architectures*, Part I: IEEE ASSP Magazine, Oct 1988. Part II: IEEE ASSP Magazine, Jan 1989

[Lee00]   C. Lee: *MediaBench*, www.cs.ucla.edu/~leec/mediabench, 2000

[LeMa95a]   R. Leupers, P. Marwedel: *Time-Constrained Code Compaction for DSPs*, 8th Int. System Synthesis Symposium (ISSS), 1995

[LeMa95b]   R. Leupers, P. Marwedel: *A BDD-based Frontend for Retargetable Compilers*, European Design & Test Conference (ED & TC), 1995

[LeMa96a]   R. Leupers, P. Marwedel: *Instruction Selection for Embedded DSPs with Complex Instructions*, European Design Automation Conference (EURO-DAC), 1996

[LeMa96b]   R. Leupers, P. Marwedel: *Algorithms for Address Assignment in DSP Code Generation*, Int. Conference on Computer-Aided Design (ICCAD), 1996

[LeMa97]   R. Leupers, P. Marwedel: *Retargetable Generation of Code Selectors from HDL Processor Models*, European Design & Test Conference (ED & TC), 1997

[LeMa98]   R. Leupers, P. Marwedel: *Retargetable Code Generation based on Structural Processor Descriptions*, Design Automation for Embedded Systems, Vol. 3, No. 1, Kluwer Academic Publishers, 1998

[LeMa99]   R. Leupers, P. Marwedel: *Function Inlining under Code Size Constraints for Embedded Processors*, Int. Conference on Computer-Aided Design (ICCAD), 1999

[Leng90]   T. Lengauer: *Combinatorial Algorithms for Integrated Circuit Layout*, Teubner/John Wiley and Sons Ltd., 1990

[Leup00a]   R. Leupers: *Register Allocation for Common Subexpressions in DSP Data Paths*, Asia South Pacific Design Automation Conference (ASP-DAC), 2000

[Leup00b]   R. Leupers: *Code Selection for Media Processors with SIMD Instructions*, Design Automation & Test in Europe (DATE), 2000

[Leup97]   R. Leupers: *Retargetable Code Generation for Digital Signal Processors*, Kluwer Academic Publishers, 1997

[Leup99]   R. Leupers: *Exploiting Conditional Instructions in Code Genera-tion for Embedded VLIW Processors*, Design Automation & Test in Europe (DATE), 1999

[LeSt98]   C.G. Lee, M.G. Stoodley: *Simple Vector Microprocessors for Mul-timedia Applications*, 31st Int. Symp. on Microprogramming and Microarchitecture (MICRO-98), 1998

[Levy97]   M. Levy: *C Compilers for DSPs flex their Muscles*, EDN Access, Issue 12, www.ednmag.com, 1997

[LeWo98]   H. Lekatsas, W. Wolf: *Code Compression for Embedded Systems*, 35th Design Automation Conference (DAC), 1999

[Liem97]   C. Liem: *Retargetable Compilers for Embedded Core Processors*, Kluwer Academic Publishers, 1997

[LiYe97]   T. Lindholm, F. Yellin: *The Java Virtual Machine Specification*, Addison Wesley, 1997

[LMD94]   B. Landwehr, P. Marwedel, R. Dömer: *OSCAR: Optimum Simul-taneous Scheduling, Allocation, and Resource Binding based on Integer Programming*, European Design Automation Conference (EURO-DAC), 1994

[LMP94a]   C. Liem, T. May, P. Paulin: *Instruction-Set Matching and Selection for DSP and ASIP Code Generation*, European Design and Test Conference (ED & TC), 1994

[LMP94b]   C. Liem, T. May, P. Paulin: *Register Assignment through Re-source Classification for ASIP Microcode Generation*, Int. Conf. on Computer-Aided Design (ICCAD), 1994

[LMW99]   Y. Li, S. Malik, A. Wolfe: *Performance Estimation of Embedded Software with Instruction Cache Modeling*, ACM Trans. on Design Automation of Electronic Systems (TODAES), Vol. 4, No. 3, 1999

[LPCJ95]   C. Liem, P. Paulin, M. Cornero, A. Jerraya: *Industrial Experience Using Rule-driven Retargetable Code Generation for Multimedia Applications*, 8th Int. Symp. on System Synthesis (ISSS), 1995

[LPJ96]   C. Liem, P.Paulin, A. Jerraya: *Address Calculation for Retar-getable Compilation and Exploration of Instruction-Set Architec-tures*, 33rd Design Automation Conference (DAC), 1996

[LSM94]   R. Leupers, W. Schenk, P. Marwedel: *Retargetable Assembly Code Generation by Bootstrapping*, 7th Int. Symp. on High-Level Syn-thesis (HLSS), 1994

[LSU89]   R. Lipsett, C. Schaefer, C. Ussery: *VHDL: Hardware Description and Design*, Kluwer Academic Publishers, 1989

[LTMF97]   M. Lee, V. Tiwari, S. Malik, M. Fujita: *Power Analysis and Min-imization Techniques for Embedded DSP Software*, IEEE Trans. on VLSI Systems, Vol. 5, No. 2, 1997

[LVK+95]   D. Lanneer, J. Van Praet, A. Kifli, K. Schoofs, W. Geurts, F. Thoen, G. Goossens: *CHESS: Retargetable Code Generation for Embed-ded DSP Processors*, chapter 5 in [MaGo95]

[MaBr91]   T. Mason, D. Brown: *lex & yacc*, O'Reilly & Associates, 1991

[MaGo95]   P. Marwedel, G. Goossens (eds.): *Code Generation for Embedded Processors*, Kluwer Academic Publishers, 1995

[Marw93]   P. Marwedel: *Tree-based Mapping of Algorithms to Predefined Structures*, Int. Conf. on Computer-Aided Design (ICCAD), 1993

[Ment93]   Mentor Graphics Corporation: *DSP Architect DFL User's and Ref-erence Manual, V 8.2_6*, 1993

[MLC+92]   S.A. Mahlke, D.C. Lin, W.Y. Chen, R.E. Hank, R.A. Bringman: *Effective Compiler Support for Predicated Execution Using the Hyperblock*, 25th Int. Symp. on Microprogramming and Microar-chitecture (MICRO-92), 1992

[MLF98]   P. Marwedel, R. Leupers, G. Fettweis: *Prozessorarchitekturen und Compilertechniken zur verlustarmen digitalen Signalver-arbeitung*, Projektantrag im DFG-Schwerpunktprogramm VIVA ("Grundlagen und Verfahren verlustarmer Informationsverar-beitung"), University of Dortmund/TU Dresden, 1998

[MME90]   M. Mahmood, F. Mavaddat, M.I. Elmasry: *Experiments with an Efficient Heuristic Algorithm for Local Microcode Generation*, Int. Conf. on Computer Design (ICCD), 1990

[Morg98]   R. Morgan: *Building an Optimizing Compiler*, Butterworth-Heinemann, 1998

[MPE99]   B. Mesman, C. Alba Pinto, K. van Eijk: *Efficient Scheduling of DSP Code on Processors with Distributed Register Files*, 12th Int. Symp. on System Synthesis (ISSS), 1999

[MPEG00]   MPEG Software Simulation Group: www.mpeg.org, 2000

[Much97]   S.S. Muchnik: *Advanced Compiler Design & Implementation*, Morgan Kaufmann Publishers, 1997

[Nabe00]   M. Naberezny: www.6502.org, 2000

[Niem98]   R. Niemann: *Hardware/Software Codesign for Data Flow Domi-nated Embedded Systems*, Kluwer Academic Publishers, 1998

[NiPo91]   A. Nicolau, R. Potasman: *Incremental Tree Height Reduction for High-Level Synthesis*, 28th Design Automation Conference (DAC), 1991

[NND95]   S. Novack, A. Nicolau, N. Dutt: *A Unified Code Generation Approach using Mutation Scheduling*, chapter 12 in [MaGo95]

[Nowa87]   L. Nowak: *Graph based Retargetable Microcode Compilation in the MIMOLA Design System*, 20th Ann. Workshop on Microprogramming (MICRO-20), 1987

[Oust94]   J.K. Ousterhout: *Tcl and the Tk toolkit*, Addison-Wesley, 1994

[PCL+96]   P. Paulin, M. Cornero, C. Liem, et al.: *Trends in Embedded Systems Technology*, in: M.G. Sami, G. De Micheli (eds.): *Hardware/Software Codesign*, Kluwer Academic Publishers, 1996

[PeWi96]   A. Peleg, U. Wieser: *MMX Technology Extensions to the Intel Architecture*, IEEE Micro 16(4), 1996

[Phil00]   Philips Semiconductors: www.trimedia.philips.com, 2000

[Plau00]   P.J. Plauger: *Embedded C++: An Overview*, www.embedded.com, 2000

[PLN92]   D.B. Powell, E.A. Lee, W.C. Newman: *Direct Synthesis of Optimized DSP Assembly Code from Signal Flow Block Diagrams*, Proc. International Conference on Acoustics, Speech, and Signal Processing, 1992

[Plum00]   Plum Hall Inc., www.plumhall.com, 2000

[PSLM+98]   C. Passerone, C. Sansoe, L. Lavagno, R. McGeer, J. Martin, P. Passerone, A. Sangiovanni-Vincentelli: *Modeling Reactive Systems in Java*, ACM Trans. on Design Automation of Electronic Systems (TODAES), Vol. 3, No. 4, 1998

[PWW97]   A. Peleg, S. Wilkie, U. Weiser: *Intel MMX for Multimedia PCs*, Comm. of the ACM, vol. 40, no. 1, 1997

[RAJ99]   P. Ranganathan, S. Adve, N.P. Jouppi: *Performance of Image and Video Processing with General-Purpose Processors and Media ISA Extensions*, 26th Int. Symp. on Computer Architecture, 1999

[RaPa99]   A. Rao, S. Pande: *Storage Assignment using Expression Tree Transformations to Generate Compact and Efficient DSP Code*, ACM SIGPLAN Conference on Programming Language Design and Implementation (PLDI), 1999

[RaPe96]   J.Rabaey, M. Pedram (eds.): *Low Power Design Methodologies*, Kluwer Academic Publishers, 1996

[RiHi88]   K. Rimey, P.N. Hilfinger: *Lazy Data Routing and Greedy Scheduling for Application-Specific Signal Processors*, 21st Annual Workshop on Microprogramming and Microarchitecture (MICRO-21), 1988

[RJD98]   A. Raghunathan, N. Jha, S. Dey: *High-Level Power Analysis and Optimization*, Kluwer Academic Publishers, 1998

[RKA99]   B. Rau, V. Kathail, S. Aditya: *Machine Description Driven Compilers for EPIC and VLIW Processors*, Design Automation for Embedded Systems, Vol. 4, No. 2/3, Kluwer Academic Publishers, 1999

[RLW99]   A. Ropers, H.W. Löllmann, J. Wellhausen: *DSPStone TI TMS320C54x*, Technical Report IB 315 1999/9-ISS, Dept. of Electrical Engineering, Institute for Integrated Systems for Signal Processing, University of Aachen, Germany, 1999

[RoFe98]   A. Römer, G. Fettweis: *Neuer Ansatz für die Codegenerierung mit Hilfe des Viterbi-Algorithmus*, DSP Deutschland, 1998

[RoWe97]   W. Rosenstiel, C. Weiler: *Using Java in Embedded Systems Design*, Proc. SASIMI, 1997

[RRD99]   K. Roy, A. Raghunatan, S. Dey: *Low-Power Design Methodologies for Systems-on-Chips*, Tutorial, 12th Int. Conf. on VLSI Design, 1999

[Sand95]   G. Sander: *VCG – Visualization of Compiler Graphs*, User Documentation V 1.30, Technical Report, Dept. of Computer Science, University of Saarland, Germany, 1995, software available via ftp.cs.uni-sb.de/pub/graphics/vcg

[SC00]   Open SystemC Initiative: www.systemc.org, 2000

[Scho99]   M. Schölzel: *Dokumentation zum Gepard C Compiler*, Technical Report, BTU Cottbus, 1999

[SCL96]   M. Saghir, P. Chow, C. Lee: *Exploiting Dual Data-Memory Banks in Digital Signal Processors*, 7th International Conference on Architectural Support for Programming Languages and Operating Systems, 1996

[SCOP99]   4th International Workshop on Software and Compilers for Embedded Systems (SCOPES '99): ls12-www.cs.uni-dortmund.de/scopes-99, 1999

[SIA98]   Semiconductor Industry Association: *International Technology Roadmap for Semiconductors*, www.semichips.org, 1998

[SLD97]   A. Sudarsanam, S. Liao, S. Devadas: *Analysis and Evaluation of Address Arithmetic Capabilities in Custom DSP Architectures*, Design Automation Conference (DAC), 1997

[SMIN96]   N. Sugino, H. Miyazaki, S. Iimuro, A. Nishihara: *Improved Code Optimization Method Utilizing Memory Addressing Operations* and its Application to DSP Compilers, Int. Symp. on Circuits and Systems (ISCAS), 1996

[SMN97]   N. Sugino, H. Miyazaki, A. Nishihara: *DSP Code Optimization Methods Utilizing Addressing Operations at the Codes without Memory Accesses*, IEICE Trans. Fundamentals, vol. E80-A, no. 12, 1997

[SMT+95] M. Strik, J. van Meerbergen, A. Timmer, J. Jess, S. Note: *Efficient Code Generation for In-House DSP Cores*, European Design and Test Conference (ED & TC), 1995

[Stro87]   B. Stroustrup: *The C++ Programming Language*, Addison-Wesley, 1987

[Suda98]   A. Sudarsanam: *Code Optimization Libraries for Retargetable Compilation for Embedded Digital Signal Processors*, Ph.D. thesis, Princeton University, Department of Electrical Engineering, 1998

[SUIF00]   The Stanford Compiler Group: suif.stanford.edu, 2000

[SuMa95] A. Sudarsanam, S. Malik: *Memory Bank and Register Allocation in Software Synthesis for ASIPs*, Int. Conf. on Computer-Aided Design (ICCAD), 1995

[Sun00]   Sun Microsystems: java.sun.com/products/embeddedjava, 2000

[Syst00]   Systemonic AG: www.systemonic.com, 2000

[Targ00]   Target Compiler Technologies: www.retarget.com, 2000

[Tech00]   Texas Instruments: *TI DSP based System-on-a-Chip Roadmap*, TI Technology Innovations, vol. 2, Jan 2000

[Tens00]   Tensilica Inc.: www.tensilica.com, 2000

[ThMo91] D.E. Thomas, P. Moorby: *The Verilog Hardware Description Language*, Kluwer Academic Publishers, 1991

[TI00]   Texas Instruments: www.ti.com/sc/c6x, 2000

[Trim00]   Trimaran – An Infrastructure for Research in Instruction-Level Parallelism, www.trimaran.org, 2000

[TSMJ95] A. Timmer, M. Strik, J. van Meerbergen, J. Jess: *Conflict Modelling and Instruction Scheduling in Code Generation for In-House DSP Cores*, 32nd Design Automation Conference (DAC), 1995

[WeGo97a] B. Wess, M. Gotschlich: *Optimal DSP Memory Layout Generation as a Quadratic Assignment Problem*, Int. Symp. on Circuits and Systems (ISCAS), 1997

[WeGo97b] B. Wess, M. Gotschlich: *Constructing Memory Layouts for Address Generation Units Supporting Offset 2 Access*, Proc. ICASSP, 1997

[Wess91] B. Wess: *Automatic Code Generation for Integrated Digital Signal Processors*, IEEE Int. Symp. on Circuits and Systems (ISCAS), 1991

[Wess92] B. Wess: *Automatic Instruction Code Generation based on Trellis Diagrams*, IEEE Int. Symp. on Circuits and Systems (ISCAS), 1992

[Wess95] B. Wess: *Code Generation Based on Trellis Diagrams*, chapter 11 in [MaGo95]

[Wess00] B.Wess: *Simulated Evolutionary Code Generation for Heterogeneous Memory-Register DSP Architectures*, European Signal Processing Conference (EUSIPCO), 2000

[WFL+99] M. Weiss, G. Fettweis, M. Lorenz, R. Leupers, P.Marwedel: *Toolumgebung für plattformbasierte DSPs der nächsten Generation*, DSP Deutschland, 1999

[WGHB94] T. Wilson, G. Grewal, B. Halley, D. Banerji: *An Integrated Approach to Retargetable Code Generation*, 7th Int. Symp. on High-Level Synthesis (HLSS), 1994

[Wiel00] R. van de Wiel: *Code Compaction Bibliography*, www.win.tue.nl/~rikvdw/bibl.html, 2000

[WiMa95] R. Wilhelm, D. Maurer: *Compiler Design*, Addison-Wesley, 1995

[YXI00] Y Explorations Inc.: www.yxi.com, 2000

[ZhGa99] J. Zhu, D. Gajski: *A Retargetable Ultra-Fast Instruction Set Simulator*, Design, Automation and Test in Europe (DATE), 1999

[ZTM95] V. Zivojnovic, S. Tjiang, H. Meyr: *Compiled Simulation of Programmable DSP Architectures*, IEEE Workshop on VLSI Signal Processing, 1995

[ZVSM94] V. Zivojnovic, J.M. Velarde, C. Schläger, H. Meyr: *DSPStone – A DSP-oriented Benchmarking Methodology*, Int. Conf. on Signal Processing Applications and Technology (ICSPAT), 1994

About the Author

 **Dr. Rainer Leupers** is a senior researcher and lecturer at the Department of Computer Science of the University of Dortmund, Germany. He obtained the Diploma and Ph.D. degrees in Computer Science with distinction from the University of Dortmund in 1992 and 1997, respectively, where he specialized in computer engineering and VLSI CAD. He received a scholarship from Siemens AG and awards for outstanding theses. Since 1993, he has been working with Prof. Peter Marwedel as a member of the Embedded Systems Group at Dortmund, where he is responsible for research projects in the area of compilers for embedded processors. Dr. Leupers has been a co-organizer of the SCOPES (Software and Compilers for Embedded Systems) workshop series, and he serves in the program committees of several design automation, DSP, and compiler conferences. In 1997, he authored the book "Retargetable Code Generation for Digital Signal Processors", published by Kluwer. In addition to his research and teaching activities, Dr. Leupers is responsible for commercial compiler development projects at the ICD technology transfer company at Dortmund.

Index

access graph, 34
address generation unit, 30
address register, 30
AGU, 30
AGU parameters, 31
array access optimization, 46
ASIP, 6, 14
assembly optimizer, 88
attribute grammar, 173
auto-increment, 31
auto-modify, 31

backend, 165, 178
basic block, 61
behavioral synthesis, 1
benchmarking, 11
branch-and-bound, 54, 153

C language, 8
C++ language, 8
C-INS scheme, 131
C-JMP scheme, 131
C-to-C transformation, 172
call graph, 152
chained instruction, 64
chromosome, 39
CISC, 4
clustered VLIW, 81
code compression, 11, 82
code generator generator, 13
code selection, 17
code size optimization, 11
common subexpression, 62
common subexpression elimination, 16
compilation speed, 15
compiler intrinsics, 106
compiler-known function, 106
conditional instruction, 6
conditional instructions, 127
constant folding, 16

constraint logic programming, 23
control hazard, 128
copy operation, 83
core, 2
correctness, 12
cross path, 84
crossover, 39
CSE, 62

data flow graph, 62
data flow tree, 62
dead code elimination, 16
def-use chain, 170
delay slot, 84, 128
DFG, 62
DFL language, 9
DFT, 62
distance graph, 50
DSP, 5
dynamic calls, 150
dynamic programming, 67, 139

embedded processor, 4
embedded system, 2
extended distance graph, 52

fitness function, 39
floating frame pointer, 33
frontend, 165, 172
function call overhead, 150
function inlining, 149

genetic algorithm, 39
graph coloring, 19
graph matching, 50

halting problem, 10
Hamiltonian path, 35
hardware-software codesign, 3

215